Why Artificial Intelligence? AI Tools in Information Technology and Foreign Language Education at the Tertiary Level

Why Artificial Intelligence? AI Tools in Information Technology and Foreign Language Education at the Tertiary Level

Natalia Shumeiko
Kateryna Osadcha
Mária Spišiaková

PETER LANG

Berlin · Bruxelles · Chennai · Lausanne · New York · Oxford

Bibliographic Information published by the Deutsche Nationalbibliothek
The Deutsche Nationalbibliothek lists this publication in the Deutsche Nationalbibliografie; detailed bibliographic data is available online at http://dnb.d-nb.de.

Library of Congress Cataloging-in-Publication Data
Library of Congress Control Number: 2025035838

Cover image: Conny Schneider on unsplash.com
Cover design by Peter Lang Group AG
This monograph is a part of the KEGA Project 012EU-4/2023 entitled 'ePortfolio as Pedagogy Facilitating Integrative Learning' supported by the Ministry of Education, Research, Development and Youth of the Slovak Republic
The research is funded by the EU NextGenerationEU through the Recovery and Resilience Plan for Slovakia under the project No. 09I03-03-V01-00045.

ISBN	978-3-631-92711-3 (Print)
ISBN	978-3-631-93872-0 (ePDF)
ISBN	978-3-631-93873-7 (ePUB)
DOI	10.3726/b23215

© 2025 Peter Lang Group AG, Lausanne (Switzerland)
Published by Peter Lang GmbH, Berlin (Germany)

info@peterlang.com

All rights reserved.

All parts of this publication are protected by copyright.
Any utilization outside the strict limits of the copyright law, without the permission of the publisher, is forbidden and liable to prosecution.
This applies in particular to reproductions, translations, microfilming, and storage and processing in electronic retrieval systems.

This monograph has undergone double-blind peer review.

An open peer review was conducted by the reviewers:

Nataliia Valko, Doctor of Pedagogical Sciences, Professor of the Department of Computer Science and Software Engineering, Kherson State University (Ukraine)

Volodymyr Proshkin, Doctor of Pedagogical Sciences, Professor of the Department of Mathematics and Physics, Boris Grinchenko Kyiv Metropolitan University (Ukraine)

www.peterlang.com

Contents

List of Figures vii
List of Tables xiii
Preface xv
Acknowledgments xvii
List of Abbreviations xix
Introduction xxi

1 AI in Tertiary Education: A Reflection on Possibilities 1

2 Microsoft Copilot and Information Technology Education 29

3 AI and Foreign Language Education: The Theoretical and Practical Dimensions 129

Conclusion 193
Appendix A 195
Appendix B 199
Appendix C 203
Appendix D 207

List of Figures

Figure 1.1 The Scopus AI concept map generated considering the set question … 10

Figure 1.2 The number of publications in Scopus (by year) … 11

Figure 1.3 AI technologies and AI tools described in scientific papers selected from the search results in Scopus AI … 12

Figure 1.4 The purposes of using artificial intelligence in higher education … 13

Figure 1.5 The relationship between types of bots in Google Trends (Source: <https://trends.google.com/trends/explore?date=all&q=Utility%20bots,Sociable%20bots,Chatbots,Assistant%20%20bots&hl=uk>). … 14

Figure 1.6 The relationship between the term "AI chatbots" and the term "Chatbots" (Source: <https://trends.google.com/trends/explore?date=all&q=Chatbots,AI%20chatbots&hl=uk>). … 15

Figure 1.7 The best AI chatbots (data from Rebelo, 2023) … 16

FIGURES

Figure 2.1 A screenshot from the website: <copilot.microsoft.com> ... 30

Figure 2.2 A sidebar of Microsoft Copilot's browser ... 33

Figure 2.3 Switching to a full-window chat mode ... 34

Figure 2.4 A full-window chat mode ... 34

Figure 2.5 Switching between a chat page and a traditional search page of Copilot ... 35

Figure 2.6 An icon of Copilot's app ... 36

Figure 2.7 A Copilot app (Android) ... 37

Figure 2.8 A query field of Copilot's app ... 38

Figure 2.9 A text query and a response from Copilot ... 39

Figure 2.10 A request to Copilot for an image generation and its result ... 40

Figure 2.11 The use of an image to "communicate" with Copilot (taking a photo) ... 41

Figure 2.12 The use of an image to "communicate" with Copilot (downloading an image) ... 42

Figure 2.13 A voice input in Copilot ... 43

Figure 2.14 The range of possible actions (i.e., copy, share, good response) in Copilot ... 44

Figure 2.15 A prompt box of the MS Copilot Chat ... 45

Figure 2.16 The answer of Copilot to the user's question ... 46

Figure 2.17 Creation of a voice request in the MS Copilot Chat 46

Figure 2.18 The icon in the input bar of Copilot 47

Figure 2.19 The answer of Copilot to the user's question 48

Figure 2.20 The follow-up questions generated by Copilot 49

Figure 2.21 The response of Copilot to one of the follow-up questions 49

Figure 2.22 The Copilot's analysis of the file (Guidelines Conference. doc) uploaded by the user 50

Figure 2.23 Creating a summary of the web page 51

Figure 2.24 A plan of the route (compiled by Copilot) 60

Figure 2.25 The plan of the route (compiled by Copilot) 63

Figure 2.26 A list of references (compiled by Copilot) 108

Figure 2.27 Tic-tac-toe (compiled by Copilot) 112

Figure 2.28 Trivia (compiled by Copilot) 113

Figure 2.29 Copilot generates an image 114

Figure 2.30 Bing Image Creator generates an image 115

Figure 2.31 Bing Image Creator using "boost" or "boosted generation" (Figure 2.31). 116

Figure 2.32 "Boost" in Bing Image Creator 117

Figure 2.33 Three-panel comic strip that shows Alan Turing breaking ENIGMA 117

FIGURES

Figure 2.34 Three-panel comic strip that shows Alan Turing breaking ENIGMA (a description) 118

Figure 2.35 Logo (compiled by Bing Image Creator) 118

Figure 2.36 Python code 119

Figure 3.1 AWCF model (in Barrot, 2021, Long, 1996) 144

Figure 3.2 Axial codes for speaking (Source: Crompton et al., 2024) 147

Figure 3.3 Axial codes for writing (Source: Crompton et al., 2024) 148

Figure 3.4 Axial codes for reading (Source: Crompton et al., 2024) 149

Figure 3.5 Axial codes for pedagogy (Source: Crompton et al., 2024) 149

Figure 3.6 Axial codes for self-regulation (Source: Crompton et al., 2024) 150

Figure 3.7 Challenges with the use of AI in ELT/L (Source: Crompton et al., 2024) 151

Figure 3.8 Speechify <www.speechify.com>: key characteristics 153

Figure 3.9 Speechify: questions for learners 154

Figure 3.10 Speechify: the possibility to choose the speaker by selecting the voice that the user likes 156

Figure 3.11 ELSA <www.elsaspeak.com>: possibilities for learners 157

Figure 3.12 ELSA: pronunciation of the word and real-time feedback 158

Figure 3.13 ELSA: pronunciation of the sentence and real-time feedback 159

Figure 3.14 Communication with ELSA with real-time feedback ... 160

Figure 3.15 Leya—AI English tutor: compelling benefits for users (Source: <leyaai.com>) ... 162

Figure 3.16 Loora—an AI English tutor: service for the business team (Source: loora.ai) ... 162

Figure 3.17 CATHOVEN Language Hub: a multiple-choice task (Analysis by <nexthub.cathoven.com>); Appendix A—questions/answers. ... 164

Figure 3.18 CATHOVEN Language Hub: the download options and the option for copying a multiple-choice task (analysis by <nexthub.cathoven.com>) ... 165

Figure 3.19 CATHOVEN Language Hub: A True/False/Not given exercise (Analysis by <nexthub.cathoven.com>); Appendix B—questions ... 166

Figure 3.20 CATHOVEN Language Hub: Providing feedback on students' writing (Analysis by <nexthub.cathoven.com>); Appendix C—feedback ... 167

Figure 3.21 The use of the method of analysis and synthesis for the determination of the reasons why students of economics learn the first foreign language ... 170

Figure 3.22 Reasons why students of economics learn English as the first foreign language ... 171

Figure 3.23 Reasons why students of economics learn the second (German, Spanish, French, Russian, or Italian) foreign language ... 172

Figure 3.24 Reasons why students of economics learn a third (Spanish, French, or Italian) foreign language ... 172

xi

FIGURES

Figure 3.25 Benefits from knowledge of foreign languages: students' views 174

Figure 3.26 Plans for the use of the knowledge of foreign languages in the professional activity 174

Figure 3.27 The development of speaking and writing skills, grammar competence, and language comprehension in class 175

Figure 3.28 The pedagogical (instructional) strategy for the implementation of AI-related technologies in teacher-led language education of students of economics 180

Figure 3.29 Subskill—vocabulary—in Axial codes for reading (Source: Crompton et al., 2024) 182

Figure 3.30 Subskill—vocabulary learning—in Axial codes for writing (Source: Crompton et al., 2024) 182

Figure 3.31 Vocab-Expander: specialized vocabulary in tabular form (Source: <vocab-expander.com>) 184

Figure 3.32 Subskill—grammar—in Axial codes for writing (Source: Crompton et al., 2024) 184

Figure 3.33 Subskill—translation—in Axial codes for writing (Source: Crompton et al., 2024) 184

Figure 3.34 Subskill—digital translation—in Axial codes for reading (Source: Crompton et al., 2024) 184

List of Tables

Table 1.1 Key findings, best practices, key points from publications on the research topic ... 2

Table 1.2 The best AI chatbots (data from Abdullahi, 2023) ... 18

Table 1.3 The best AI chatbots on the market (data from Stefanowicz, 2024) ... 21

Table 1.4 The synthesis based on and indicated in three rankings of the best and the most popular AI chatbots ... 22

Table 1.5 The top five best/the most popular AI chatbots based on the conducted analysis of 3 ratings ... 23

Table 3.1 Key findings, best practices, key points from publications on the research on AI in language education ... 131

Table 3.2 Benefits (affordances) and challenges of using AI-powered systems and applications in ELT ... 145

Table 3.3 A comparison table of options provided by Leya, an AI English tutor, with human tutor work ... 161

Table 3.4 Weaknesses and strengths of AI apps ... 163

Table 3.5 Evaluation of how well the topics that students study and discuss in foreign language classes prepare them for their future professional careers and contribute to the development of their knowledge and skills (the assessment of the significance in the order of importance by way of an eight-point Likert scale from 1 [extremely relevant] to 8 [not at all relevant]) ... 178

Table 3.6 How difficult it is to acquire knowledge and skills to improve the level of foreign languages (the assessment of the significance in the order of importance by way of a nine-point Likert scale from 1 [extremely difficult] to 9 [not at all difficult]) ... 178

Table 3.7 What competencies and skills do you consider very well-formed? (the assessment of the significance in the order of importance by way of a nine-point Likert scale from 1 [extremely well-formed] to 9 [not well-formed]) ... 179

Table 3.8 AI tutorial systems for grammar correction ... 183

Preface

AI has emerged as a transformative force in tertiary education, reshaping both pedagogical theory and classroom practice. Against the backdrop of rapid technological change, the authors of this monograph critically examine the integration of the latest AI tools and technology, such as Microsoft Copilot, ChatGPT, DALL-E, the Cathoven Language Hub, ELSA (an AI-powered English-speaking coach platform), AI-powered English language tutors, and Speechify (a text-to-speech technology) in the domains of IT education and foreign language learning.

Chapter 1 situates the authors' exploration of the scientific publications related to the research topic. The authors performed a bibliometric analysis of scientific works, thereby representing existing research trends and suggesting the perspectives of leading scholars on AI in education. This chapter considers the role of AI chatbots in training students at the tertiary level and provides an evaluation point of view on their pedagogical value in the contemporary educational environment.

Chapter 2 focuses on the use of GenAI in IT education, with a particular emphasis on Microsoft Copilot. The authors concentrate attention on the benefits and opportunities that Microsoft Copilot provides. Prompts for Microsoft Copilot and practical use cases are presented in this chapter of the monograph. The potential challenges of integrating AI-generated content into tertiary education curricula are outlined.

Chapter 3 explores the integration of AI into foreign language education. Based on a thorough analysis, the authors outline the characteristics of

AI-driven language-learning and teaching technology, focusing on English language tutors (Leya <leyaai.com>, Loora <loora.ai>) and a coach platform – the Cathoven Language Hub, detailing the technological aspects relevant to their use in pedagogical practice. A special focus is on the opportunities that AI brings to language education, supporting and strengthening foreign language acquisition, and enhancing communicative competence.

This monograph aims to provide a clear and comprehensive description of AI tools and characterize their value for university lecturers. The work particularly focuses on fostering learners' speaking skills at the intersection of technology and education. Through the lens of the conducted theoretical analysis, combined with practical recommendations for AI use in tertiary education, the study seeks to shed light on the value of AI for the educational process at the tertiary level and outline the prospects for the future of education, equipping lecturers with the information essential for effective and appropriate AI use.

Acknowledgments

This monograph is a scientific work prepared within the research project KEGA/ePortfolio as Pedagogy Facilitating Integrative Learning (012EU-4/2023), funded by the Ministry of Education, Research, Development and Youth of the Slovak Republic. The research was supported by the European Union through the NextGenerationEU programme under the Recovery and Resilience Plan for Slovakia, Project No. 09I03-03-V01-00045.

Abbreviations

AI	Artificial Intelligence
AIEd	AI Applications in Education
AI HLEG	High-Level Expert Group on Artificial Intelligence
AR	Augmented Reality
ASR	Automatic Speech Recognition
ASLL	Autonomous Second Language Learning
AWCF	Automated Written Corrective Feedback
AWE	Automated Writing Evaluation
CALL	Computer-Assisted Language Learning
CDA	Computerized Dynamic Assessment
CDRT	Computerized Dynamic Reading Test
CEFR	The Common European Framework of Reference for Languages
CLIL	Content and Language-Integrated Learning
DA	Dynamic Assessment
DDL	Data-Driven Learning
DIALANG	A Diagnostic Language Assessment System
EBMT	Example-Based Machine Translation
EFL	English as a Foreign Language
ELT	English Language Teaching
EU	European Union
FLA	Foreign Language Anxiety
GenAI	Generative Artificial Intelligence
GPT	Generative Pre-Training Transformer

ABBREVIATIONS

HEIs	Higher education institutions
IPAs	Intelligent Personal Assistants (e.g., Siri, Google Assistant)
ITSs	Intelligent Tutoring Systems
IT	Information Technology
L2	Second Language
LA	Learner Autonomy
LGC approach	Learner-Generated Context approach
LGC	Learner-Generated Context-Based approach
LLMs	Large Language Models
MyET	My English Tutor
MT	Machine Translation
NLP	Natural Language Processing
NLU	Natural Language Understanding
NLG	Natural Language Generation
NMT	Neutral Machine Translation
OERs	Open Educational Resources
RBMT	Rule-Based Machine Translation
SCT	Sociocultural Theory
SMT	Statistical Machine Translation
T-NLG	Turing Natural Language Generation
VR	Virtual Reality
ZPD	Zone of Proximal Development

Introduction

It is an indisputable fact that AI has changed the professional landscape of tertiary education today, necessitating new approaches to training specialists, including IT professionals, for all business spheres. AI continues to shape our world today. With the advent of generative AI, primarily, questions related to the ways of teaching arose. However, questions related to assessments, selecting relevant and truthful information for teaching students, and academic integrity are also valid concerns for many educators today. Gradually, other issues have become important today; we mean the problems closely linked to ethical concerns such as personal data protection. The unique opportunities that AI technology provides are known worldwide. The risks, among which are security risks and dependence on AI, that people face while using AI resources are also evident. Educators have realized this fact. Moreover, pedagogues need to rethink the training methods and approaches that are applied in lessons, consider what students should learn, and, based on that, integrate AI into the teaching-learning process in existing educational institutions.

AI becomes a new impetus for revising and adapting curriculum content, considering best practices, adjusting teaching methods, prioritizing the development of critical thinking, and emphasizing students' ability to analyze information from many different sources in the surrounding world. Pedagogues should equip students with the competencies needed to use AI resources properly and encourage them to avoid unthinkingly relying on generated information or data. Students should also be able to work with AI tools, being aware of their opportunities and limitations. Currently, search engines, open information resources, and AI tools are a means of professional training for future specialists at the university level. Mirroring this, the roles of educators in tertiary education institutions are gradually changing due to the use of AI in education.

The monograph has moved professors and university lecturers to recognize significant issues related to AI and its application at the tertiary level. The following research questions (RQ) are defined for the study:

- RQ1: How can the potential of AI be exploited at the tertiary level to foster students' competencies, improve knowledge, and skills?
- RQ2: Can AI function as a reasonable means of education for IT students and foreign language education for students of economics?

To address these questions, three research objectives (RO) are formulated:

- RO1. To analyze knowledge (publications, views, thoughts) about AI, based on the information obtained from a bibliometric analysis of scientific publications.
- RO2. To propose pedagogically sound ways (techniques) of using AI tools in teaching foreign languages and training IT at tertiary education institutions.
- RO3. To provide examples of integrating AI technologies into the training of students (with particular emphasis on IT education and language education).

These objectives are developed in three chapters:

- Chapter 1. AI in Tertiary Education: A Reflection on Possibilities
- Chapter 2. Microsoft Copilot and Information Technology Education
- Chapter 3. AI and Foreign Language Education: The Theoretical and Practical Dimensions.

CHAPTER 1

AI in Tertiary Education: A Reflection on Possibilities

In the context of tertiary education, AI presents both opportunities and challenges. Recent developments in AI highlight the potential for reshaping teaching-learning process. Recently published papers are devoted to the relevance of applying AI resources in tertiary education. AI has immense potential to transform the teaching, learning, and organization of the educational process at the university level.

From a pedagogical perspective, AI offers new tools for students' engagement. A bibliometric analysis of scientific publications on AI in tertiary education confirmed this view. Articles retrieved from the Google Scholar search engine over the past five years using keywords (AI in higher education, AI at universities) were analyzed. To publications addressing tertiary education-related issues, special attention was given, while the authors excluded those focused on secondary and primary education.

The search criteria encompassed a range of keywords related to AI and higher education, including "Artificial Intelligence technology," "AI tools," and "Higher education." By analyzing Google Scholar data, we identified and analyzed scientific publications published from 2019 to 2024 that explored the use of AI in higher educational institutions.

Chapter 1 represents the data from a bibliometric analysis in tables and graphs. Ideas, thoughts, and concepts from approximately 100 recently published scientific works were drawn up.

1.1. A Bibliometric analysis of scientific publications on the research topic

The results of the bibliometric analysis of scientific publications in the Google Scholar search engine over the past five years on search inquiries "Artificial Intelligence in Higher Education" were conducted (39,700 scientific publications, excluding citations). We focus on those publications that shed light on higher education issues (contained words "Artificial Intelligence," "Artificial Intelligence technology," "Artificial Intelligence tools," and "AI tools"), as in the search results, we found the publications on secondary or primary education. The search results based on Google Academy data allowed distinguishing the scientific publications in the set period (2019–2023).

A detailed review of the Google Academy search results allowed us to highlight 30 scientific publications for a certain period, which discuss the issue of AI application in higher education. We determined the results of these publications by considering the titles and the abstracts of the scientific articles. If it was not clear from the title and the abstract of the scientific publication, we considered the "Conclusion" sections of the publications, or we have read and analyzed the entire content of the scientific publication (Table 1.1).

Table 1.1 Key findings, best practices, key points from publications on the research topic

No.	The author(s)	Key findings, best practices, key points
1	Zawacki-Richter et al. (2019)	Four areas of AI in Education applications in academic support services, and institutional and administrative services: 1) profiling and prediction, 2) assessment and evaluation, 3) adaptive systems and personalization, 4) intelligent tutoring systems.
2	Bates et al. (2020)	"Breakthrough" AI applications for teaching and learning are unlikely to emerge from within mainstream higher education. They are more likely to arrive from outside the formal postsecondary system, through organizations such as LinkedIn, lynda.com, Amazon or Coursera […]. The key question then is whether technology should aim to replace teachers and instructors through automation, or whether technology should be used to empower not only teachers but also learners.

No.	The author(s)	Key findings, best practices, key points
3	Hinojo-Lucena et al. (2019)	AI applied to higher education is a reality, since it is currently experimented, and beneficial results are being obtained. However, at the same time, it is a marginal reality, since it is not developed enough, and its application is not widespread.
4	Chatterjee and Bhattacharjee (2020)	In the article, the authors provided a model identifying the determinants (Perceived Risk, Effort Expectancy, Facilitating Conditions, Behavioral Intention) that would help and accelerate the adoption of AI in higher education. They have mentioned that the institutes of higher education would enjoy effective advantages if they use AI.
5	Ouyang et al. (2022)	1) the functions of AI applications in online higher education include prediction of learning status, performance or satisfaction, resource recommendation, automatic assessment, and improvement of learning experience; 2) traditional AI technologies are commonly adopted while more advanced techniques (e.g., genetic algorithm, deep learning) are rarely used yet; and 3) effects generated by AI applications include a high quality of AI-enabled prediction with multiple input variables, a high quality of AI-enabled recommendations based on student characteristics, an improvement of students' academic performance, and an improvement of online engagement and participation.
6	Dhawan and Batra (2020)	Implementation of artificial intelligence in higher education calls for attention and investment in upgrading the skills of the stakeholders to harness the potential of AI and its safe and ethical usage. A mandatory training program for teachers should be initiated. The benefits of the technology can be reaped in a better way if the students are introduced to such systems in the formative years itself. Preparing for future demands investment in infrastructure facilities and training programs. A public-private partnership model should be adopted to accelerate integration.
7	Ali and Abdel-Haq (2021)	Four areas of AI education applications in academic support services and institutional and administrative services were revealed, including profiling and prediction, assessment and evaluation, adaptive systems and personalization, and intelligent tutoring systems.
8	Ghnemat et al. (2022)	Using AI competency-based learning will let students achieve the course outcomes easier and faster and increase student engagement by solving real-life industrial problems in different application domains.

No.	The author(s)	Key findings, best practices, key points
9	Slimi and Carballido (2023)	Data collection, labeling, and algorithm documentation must be of the highest quality to ensure traceability and openness. Universities must study the ethical, social, and policy implications of AI to ensure responsible development and deployment. The AI ethics policies stress responsible AI development and deployment, with a focus on transparency and accountability. Making AI systems more transparent and answerable may reduce the adverse effects of displacement.
10	Mishra (2019)	Information systems are the key component of quality assurance leading to higher student satisfaction and further business growth for higher education institution. Using artificial intelligence and data analytics, HEIs can ensure compliance, internal quality of the educational programs, enhance student satisfaction, etc.
11	O'Dea and O'Dea (2023)	At the national level, the government should consider including AI in education in national initiatives. At the institutional level, university senior management teams should provide an environment to enable AI to operate both in terms of support and funding for appropriate technical environments and in terms of providing the opportunities to increase the skill base of academic tutors, so that AI affordances can be identified from and within the higher education sector. At the personal level, academic tutors need to take initiative to participate actively in the CPD and other training sessions provided by their university.
12	Taneri (2020)	Intelligent automation collaborating effectively with humans is an opportunity for humans to use AI as a tool for enhancing what humans do. While the growing importance of AI in the workplace will doubtless affect what universities and colleges teach, it also has an impact on how they teach and on how students learn.
13	Pedro et al. (2019)	Six challenges are presented in this publication: The first challenge lies in developing a comprehensive view of public policy on AI for sustainable development. The second challenge is to ensure inclusion and equity for AI in education. The third challenge is to prepare teachers for an AI-powered education while preparing AI to understand education. The fourth challenge is to develop quality and inclusive data systems. The fifth challenge is to make research on AI in education significant. The sixth challenge deals with ethics and transparency in data collection, use and dissemination.

No.	The author(s)	Key findings, best practices, key points
14	Zhai et al. (2021)	All empirical studies reviewed presented the positive effects of AI techniques on education. However, the interview and the review paper have, respectively, surfaced the challenges or misunderstanding of AI in education. There is a need to articulate a holistic evaluation criterion to measure the effectiveness of AI in education. To ensure the validity and reliability of the evaluation, a multidimensional model should be adopted, which includes technique, pedagogical design, domain knowledge, and human factors.
15	Cope et al. (2021)	Traditional pedagogy and its assessments, and perhaps for the era in which we now live, irredeemably so. Artificial intelligence promises a new way forward for assessment and education.
16	Tao et al. (2019)	Artificial intelligence and robotics taken to the extreme contain dangers and challenges that must be considered in all areas of their application, particularly in education. The use of robots and artificial intelligence instruments can generate disconnection with emotions, students and teachers state that a robot is not imitable because it also lacks emotions.
17	Dempere et al. (2023)	Developing AI-based tools such as ChatGPT increases the likelihood of replacing human-based teaching experiences with low-cost chatbot-based interactions. This possibility may result in biased teaching and learning experiences with reduced human connection and support. Adopting AI-based technologies like ChatGPT can provide many benefits to HEIs, including increased effectiveness on student services, admissions, retention, etc., and significant enhancements to teaching and research activities. The risks involved in adopting this technology in the education sector are substantial, including sensitive issues such as privacy and accessibility concerns, unethical use, data collection, misinformation, technology overreliance, cognitive bias, replacement of human interaction, etc.

No.	The author(s)	Key findings, best practices, key points
18	Holmes et al. (2023)	The possibilities, which have been foreshadowed by the Artificial Intelligence in Education (AIED) tool: AI to support collaborative learning, AI-driven student forum monitoring, AI to support continuous assessment, AI learning companions for students, and AI teaching assistants for teachers, AIED as a research tool to further the learning sciences (i.e., in order to help us better understand learning).
19	Rudolph et al. (2023)	Generally, we advise against a policing approach (that focuses on discovering academic misconduct, such as detecting the use of ChatGPT and other AI tools). We favor an approach that builds trusting relationships with our students in a student-centric pedagogy and assessments for and as learning rather than solely assessments of learning. The principle of constructive alignment asks us to ensure that learning objectives, learning and teaching and assessments are all constructively aligned.
20	Sullivan et al. (2023)	The public discussion and university responses about ChatGPT have focused mainly on academic integrity concerns and innovative assessment design. There is potential for AI tools to enhance student success and participation from disadvantaged backgrounds. Academics and university representatives should be aware of the frames they choose to discuss when engaging with the media, as news coverage can influence social norms toward student cheating behavior and public perceptions of universities. It is important for universities to adapt and embrace the use of AI tools in a way that supports student learning and prepares them for the challenges of an increasingly digital world.
21	Salinas-Navarro et al. (2024)	We highlighted GenAI tools potential to act as agents-to-think-about the intended learning outcomes and real-world complex learning scenarios, agents-to-teach-and-learn-with to facilitate active learning experiences, agents-to-assess-learning-with authentic assessment tasks, and agents-to-learn-with experiential learning activities for authentic assessment. These findings underscore the transformative role of GenAI tools in enhancing teaching and learning efficacy and effectiveness.

No.	The author(s)	Key findings, best practices, key points
22	Sihare (2024)	The integration of AI and ML can play a crucial role in identifying causes of dropout rates among students and developing targeted solutions. By utilizing contemporary technologies and engaging all stakeholders, including public bodies, institutions, and affected parties, it is possible to mitigate dropout rates and create a more engaging and effective educational environment.
23	Forero-Corba and Bennasar (2024)	The results show that the 33 intelligent techniques extracted from the studies can be applied in the education sector to 1) detect students' academic performance early, 2) improve the educational skills of teachers, 3) facilitate the learning of students with autism spectrum disorders, 4) predict school dropout and make decisions about it, 5) improve and generate educational content, 6) close educational gaps, 7) implement AI teaching at all educational levels, 8) strengthen the information security of the educational community, 9) motivate learning through mobile devices, 10) strengthen the field of robotics, 11) improve academic and career guidance for students, 12) prevent the spread of fake news on social networks, 13) understand and reflect on the relationship between humans and machines, and 14) develop critical thinking based on computational thinking.
24	Zhao et al. (2024)	AI writing assistants have received considerable attention in recent years as a new means to enhance students' academic writing. This paper examines the use of Wordtune, an AI-powered writing assistant, by Chinese international students in higher education through interviews. Students found the rewriting options useful, especially the function to rewrite in formal language. Students self-identifying as beginners in English used all the functions, but rather indiscriminately. Students with higher-level skills used it more selectively and learned to improve their writing through examining alternative rewrites. All users wanted the function to rewrite sentences more formally to suit an academic writing style.

No.	The author(s)	Key findings, best practices, key points
25	Vinkóczi et al. (2023)	The importance of AI in education can be captured in its ability to personalize learning pathways, improve teaching methods and automate related administrative tasks. AI technologies are able to adapt to the needs of individual learners, providing personalized instruction and improving learning outcomes. AI can also help educators by automating routine tasks, allowing them to focus on individualized instruction and create a more engaging and effective learning environment.
26	Pisica et al. (2023)	Most of the perceptions regarding the pros of AI are related to the effects on the teaching–learning process, research, and the development of new skills. As for the negative aspects, socio-psychological effects and the loss of the sense of "being human," a kind of fear of the dissolution of humans as social beings, together with security and ethical aspects are among the most prominent concerns coming from academics.
27	Schneckenleitner et al. (2023)	Universities are encouraged to conduct their own research on AI to stay abreast of the latest advancements and developments. By actively engaging in research initiatives, institutions can better understand the potential of AI and its impact on higher education, allowing them to make informed decisions regarding curriculum development, resource allocation, and strategic planning. By leveraging individualization options, addressing data protection concerns, and adapting curricula due to changing job profiles, institutions can harness the power of AI to enhance the learning experience and prepare students for the future.
28	Grájeda et al. (2024)	The derivation of a Synthetic Index of Use of AI Tools applied to higher education (SIUAIT) is one of the main contributions of this paper to the scientific community, which can be utilized from the instrument by any educational unit wishing to measure students' perceptions of using AI tools. SIUAIT has 30 items to examine five key dimensions: 1) effectiveness use of AI tools, 2) effectiveness use of ChatGPT, 3) student's proficiency using AAI tools, 4) teacher's proficiency in AI and 5) advanced student skills in AI. These dimensions form a synthetic index used for comprehensive evaluation.

No.	The author(s)	Key findings, best practices, key points
29	Jafari and Keykha (2023)	Practical suggestions and policy recommendations: 1) holding empowerment courses for faculty members and students in order to improve computer capability and adaptability to prepare for the future; 2) formulating laws at the level of higher education and universities in the field of using new technologies such as Chat GPT for more effective monitoring and follow-up of violations: 3) Changing the assessment mechanisms of students and adapting and accepting new methods according to the unique structure of artificial intelligence; 4) developing universities' hardware and software infrastructure and realizing smart universities to prepare for the digital future; 5) redefining the laws related to intellectual property laws and patents according to the changes brought about by the evolution of artificial intelligence; 6) making small changes to change the norms of the academic culture to accept technological developments.
30	Segbenya et al. (2023)	It is recommended that trainers of postgraduate students, especially graduate schools in various universities, should ensure that proper education and training are provided for postgraduate students on the uses/benefits and dangers associated with AI usage for academic activities. The training and use of the various available AI software will also be helpful to postgraduate students. Furthermore, it is recommended that managers of postgraduate schools/faculties should establish AI tracking detectors to check the level of plagiarism and acceptable level, which should be made known to students. Authorities of postgraduate schools should have rules and regulations or academic policies on allowable similarity indexes in student submissions to guide students' usage of AI software among postgraduate students. Finally, it is recommended that facilitators on postgraduate programs should vary their assessment and classwork. That is, facilitators should now adopt more practical assignments/classwork and project works that require less machine or AI text generation. These forms of assessment will call for building more human relations skills for working in groups, problem-solving skills and creativity among postgraduate students.

In this study, we applied a tool based on AI—Scopus AI. This tool allows us to navigate the rich academic landscape of the Scopus platform. Scopus AI presents reliable Scopus content. Moreover, it offers to ask the question. Then, it provides the answer in several forms, such as a summary, an expanded summary, a concept map, topic experts (3 primary publications on the topic), and "a greater detail" (suggestions for additional questions on the subject). Based on the set question ("What artificial intelligence tools are used in higher education to improve it?"), we generated a conceptual map (Figure 1.1). On this map we can see:

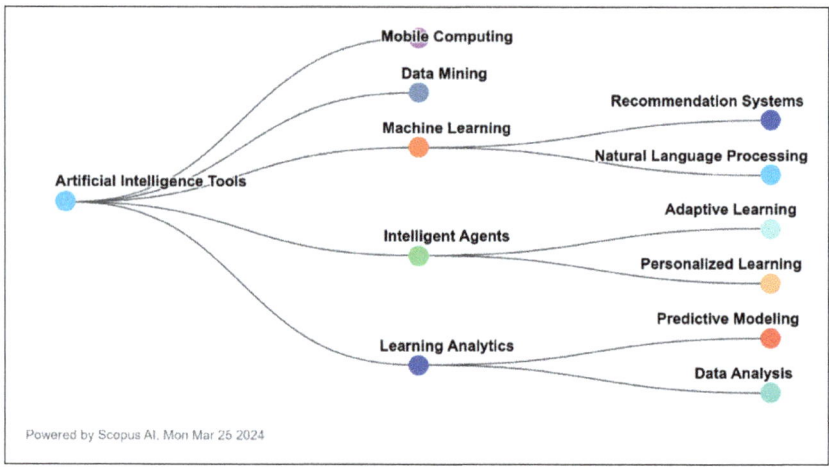

Figure 1.1 The Scopus AI concept map generated considering the set question

Analyzing the proposed conceptual map, it can be noted that the most used tools of artificial intelligence are Machine Learning, Intelligent Agents, Recommendation Systems, Natural Language Processing and Predictive Modeling.

During the study, we carried out a bibliometric analysis of scientific works in the bibliographic and abstract database of Scopus. We set the keywords: "artificial AND intelligence OR ai AND high AND education OR at AND universities." In the search process, we set the timeframe (the last five years, particularly from 2019 to 2024) and the field of science and knowledge (social sciences, computer science, engineering, multidisciplinary research, art and humanities, psychology). As a result, we received the list of 20 publications. Considering obtained results, we conclude that the highest number of articles within the set period was in 2023 (Figure 1.2).

Based on the analysis of the keywords in the selected articles we determined the key notions (AI technologies and AI tools) that are described and analyzed in them. Due to this, it is possible to determine the significance of these AI technologies and AI tools, based on the number of their mentions in the keywords of the articles under analysis (Figure 1.3).

So, according to this principle, the most often mentioned AI technologies and tools in Higher Education are Machine Learning, Deep Learning, Neural Networks, Chatbots, Convolutional Neural Network, ChatGPT, etc.

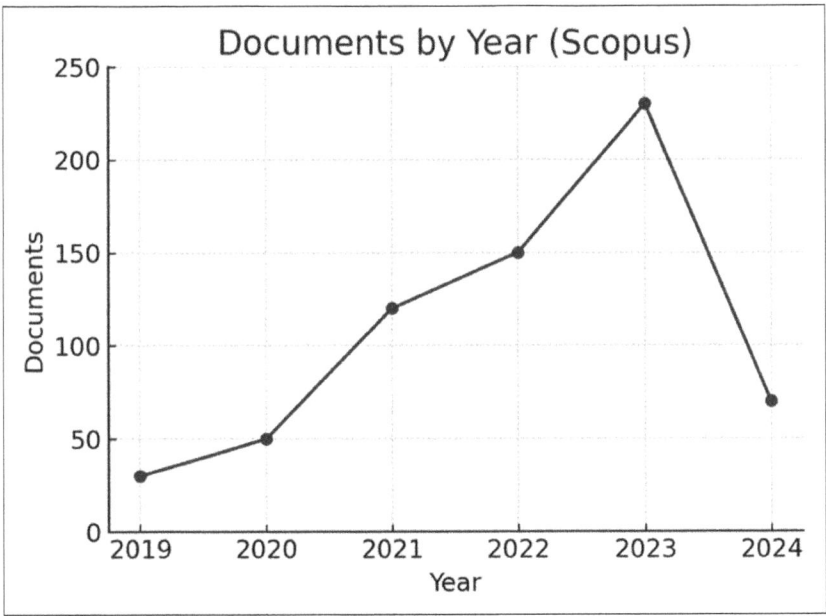

Figure 1.2 The number of publications in Scopus (by year)

CHAPTER 1

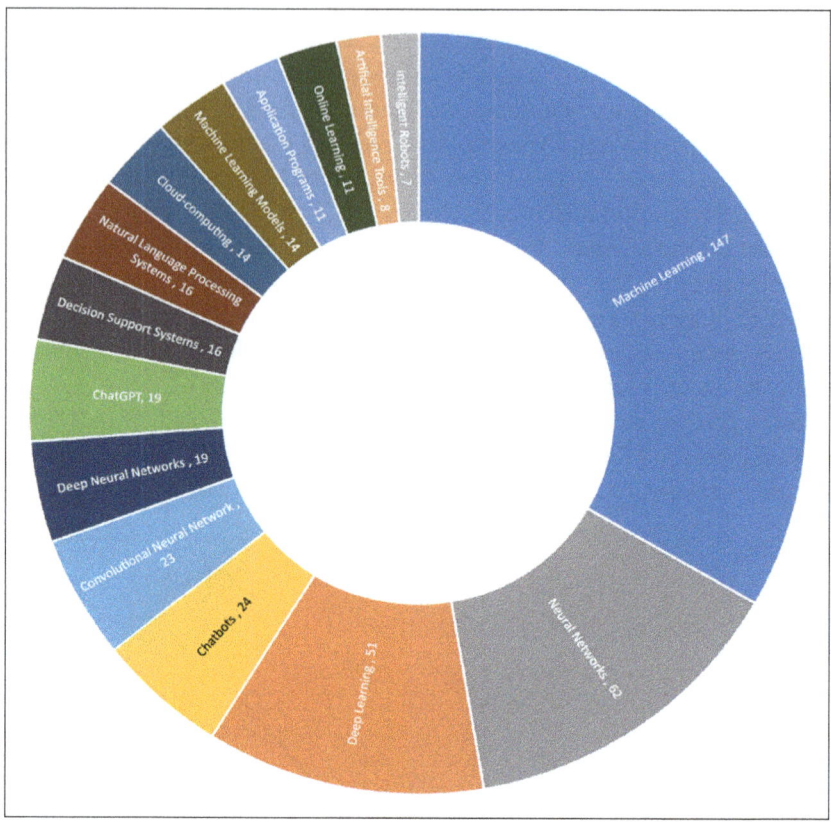

Figure 1.3 AI technologies and AI tools described in scientific papers selected from the search results in Scopus AI

Also, based on the analysis of keywords, we determined in which fields artificial intelligence is used in higher education, the results are shown in Figure 1.4.

Based on the analysis of keywords, it can be noted that artificial intelligence in higher education is most often used in the following areas: Learning Systems, E-learning, Big Data, Information, Management, Internet Of Things, Forecasting, Data Mining, Active Learning, Decision-Making, Personnel Training, etc.

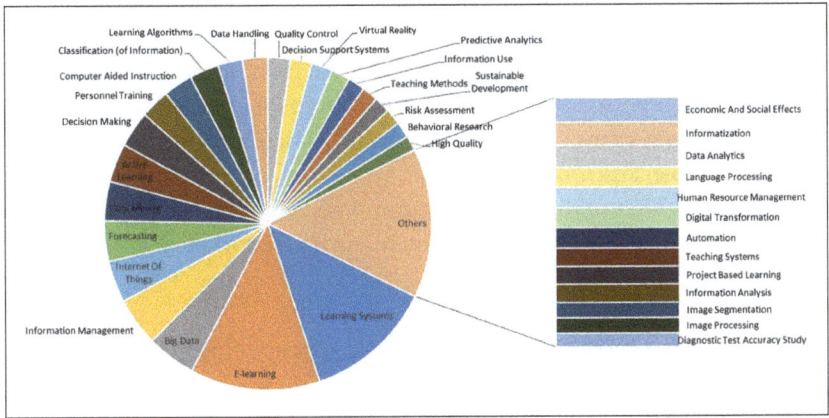

Figure 1.4 The purposes of using artificial intelligence in higher education

1.2. What is an AI Chatbot?

The public was amazed by the appearance of ChatGPT. The chat capabilities included communicating with people in natural language, writing texts at the level of copywriters, creating software code, and later generating images and analyzing data. A little later, other well-known and not-so-well-known companies began to develop their AI chats, namely, Google AI (Gemini), Bing AI Chat (Copilot), Perplexity and others. Chats also began to appear not only for generating text, but also images (DALL-E), music (Beatoven.ai) and video (Colossyan Creator).

Mariciuc (2023) defines the concept of a "bot" as a computer program launched on the internet, which is trained by a set of rules to act autonomously and offer a service. Traditionally, the terms bot and robot have been used as equivalents interchangeably, as the first is a diminutive of the second. On the market, Mariciuc highlights three categories of bots:

1. Utility bots are a type of bot which fulfills specific functions—it has a goal and is measured based on the function for which it was programmed, for example, a bot that answers frequently asked questions (FAQ).

2. Sociable bots (assistant bots) are mainly characterized by a good level of conversation because they have a language that allows them to develop their personality. They are usually present in mobile and desktop operating systems, for example, Cortana (Microsoft), Alexa (Amazon), Siri (Apple) and Google Assistant.
3. Chatbots are usually based on a computer program, capable of recognizing natural language and of maintaining a conversation with a user to solve a query.

You can trace the relationship between these types of bots in Google trends (Figure 1.5).

Based on the data analysis provided by Google Trends, Assistant Bots were popular in early 2004. However, since 2015 their popularity has been almost zero. The term Sociable bots is not found on the internet at all, and a type of bots like Utility bots was popular from 2004 to 2009. Instead, Chatbots, as a term, was trendy compared to other terms people used from 2004 to 2012. However, the term Chatbots began to regain popularity from 2016, and, later, in 2022, we have noted a sharp jump and a further increase in its popularity.

Among the variety of chatbots (simple chatbots or based on linguistic models (Rules), intelligent chatbots or based on machine learning (AI), recognition

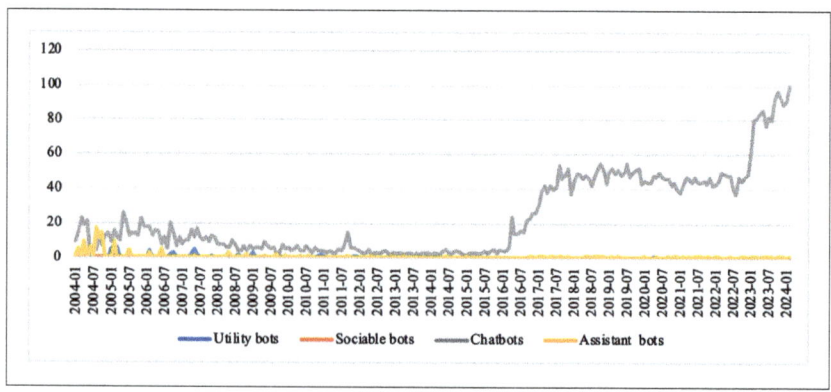

Figure 1.5 The relationship between types of bots in Google Trends (Source: <https://trends.google.com/trends/explore?date=all&q=Utility%20bots,Sociable%20bots,Chatbots,Assistant%20%20bots&hl=uk>).

chatbots or word-spotting, and cognitive chatbots), Mariciuc (2023) highlights and suggests the term "AI chatbots."

And now, using the Google Trends platform, we will trace the occurrence and popularity of the term "AI chatbots" and then compare it with the term "Chatbots" (Figure 1.6).

Based on the data analysis provided by Google Trends, the first peak of the popularity of the term "AI chatbots" was in 2004 and, later, in 2006. However, a decline subsequently began, which lasted until the end of 2016. A sharp increase in the popularity of this term and the phenomenon itself occurred at the end of 2022, and in subsequent years, we observe its constant growth.

AI chatbots are the specific software programs that simulate human-like conversation through text or speech. These bots differ from regular chatbots in that they use AI algorithms to understand the different user queries and respond appropriately. Their goal is essentially to offer information, and assistance, or engage in natural conversation. This is why they are often used in customer service and support systems across industries to automate tasks, provide guidance, or entertain users (Ventoniemi, n.d.).

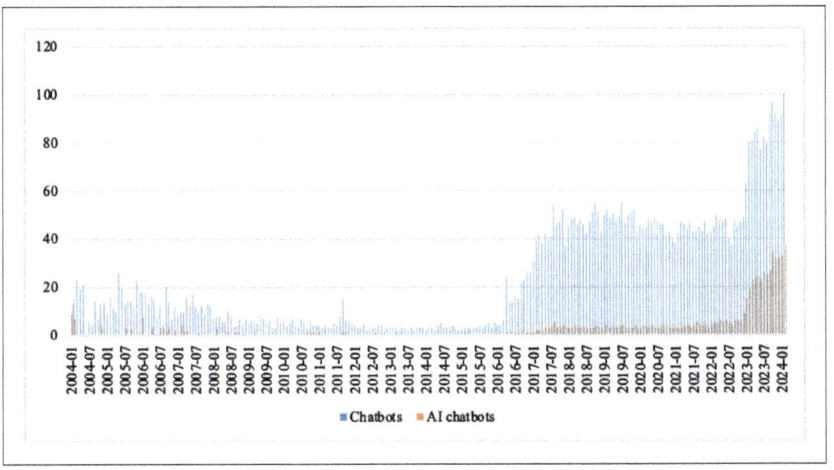

Figure 1.6 The relationship between the term "AI chatbots" and the term "Chatbots" (Source: <https://trends.google.com/trends/explore?date=all&q=Chatbots,AI%20chatbots&hl=uk>).

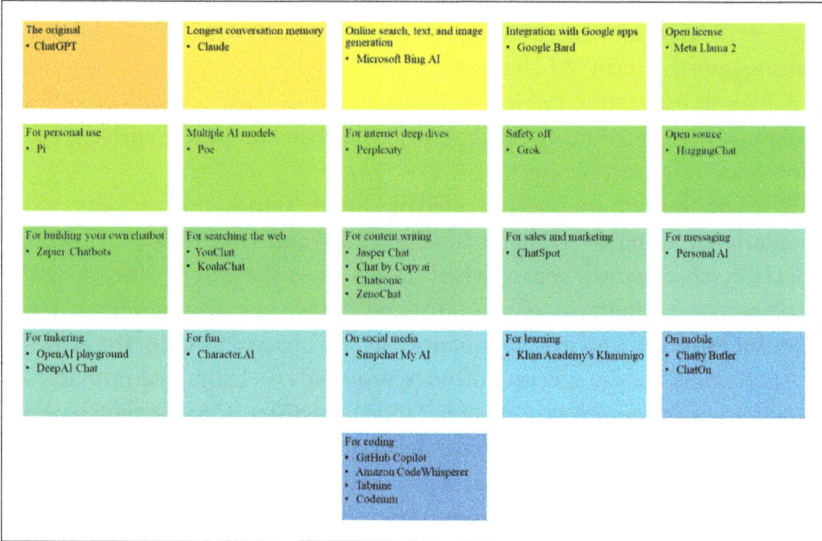

Figure 1.7 The best AI chatbots (data from Rebelo, 2023)

According to M. Rebelo, in accordance with the functions performed, it is possible to distinguish the best AI chatbots (Rebelo, 2023) (Figure 1.7).

Abdullahi (2023) classifies and distinguishes AI chatbots by the following characteristics: Best for, Query limit, Language model(s), Chrome extension created by the vendor, and Starting price. He also ranks them according to the criteria: Cost, Feature Set, Ease of Use, Quality of Output, and Support. We pooled data and summarized it in Table 1.2. Moreover, we arranged AI chatbots by rating (Table 1.2).

Stefanovich (2024) proposed a rating of chatbots on a five-point scale, taking into consideration the following criteria:

1. Easy access—The AI chatbot should not require complicated settings, be easy to use, and not require coding knowledge.
2. Accuracy—The AI chatbot should be able to give answers that have sense and are correct.

3. Chat experience—The AI chatbot should allow the user to have online conversations by exchanging messages back and forth, allowing for more natural interaction.
4. Key features—the AI chatbot should also offer additional functionality, such as availability and support of several languages, useful integrations, saving the history of conversations, copying and exporting the results, etc.

We slightly changed the appearance of the table proposed by Stefanowicz. We've ranked AI chatbots and clarified pricing information (Table 1.3).

So, let us summarize the results of 3 ratings. For this, we determine the number of repetitions of AI chatbots, remove those that are mentioned just once, and determine the overall rating by calculating the arithmetic mean of 2 ratings in numerical format (if there is no second number, then we retain only one), remove AI chatbots and the numerical ratings that are not defined (for example, Pi, Personal AI, and Snapchat My AI) (Table 1.4).

So, we determined the top five AI chatbots in each ranking of AI chatbots (Table 1.5).

To sum up, the undoubted leaders among AI chatbots are Jasper and ChatGPT. Also, the conducted analysis summarizes that Perplexity, Ada, and Google Bard have a high rating. At the same time, we notice that Jasper and Ada are used mainly for business purposes. Perplexity, Google Bard, and Microsoft Bing AI are used in education, as evidenced in numerous publications (Siegle, 2023; Rudolph, Tan, & Tan, 2023; Plevris, Papazafeiropoulos, & Jiménez, 2023). As regards the AI chatbot Replika, we should note that it was developed having in a creator's mind a clear idea to invent a personalized AI that was designed to help people express themselves and talk about themselves in a practical and frank discussion (Kourkoulou, 2023).

Thus, it is expedient to concentrate attention on these three AI chatbots: Perplexity, Google Bard, and Microsoft Bing AI. We note that the AI chatbot market is developing rapidly. The functions of different AI chatbots that affect the ratings outlined above change rapidly; therefore, such ratings are very variable. At the same time, these ratings still show trends and user preferences that allow educators to choose and use AI training tools that are the best and most appropriate for the educational processes.

Table 1.2 The best AI chatbots (data from Abdullahi, 2023)

Name	Overall rating (cost, feature set, ease of use, quality of output, support)	Best for	Query limit	Language model(s)	Chrome extension created by vendor	Starting price
Freshchat	4.63 (5, 5, 5, 5, 2.5)	Automating self-service	500 Freshbots sessions	Freddy AI, Microsoft Azure OpenAI Service	No	$23 per agent per month
Crisp Chatbot	4.55 (5, 3.5, 5, 5, 5)	Lead nurturing	Unlimited	Proprietary LLM model	No	$25 per month per workspace
ChatGPT	4.51 (4.15, 4.4, 5, 3.75, 5)	Versatility and advanced chat generative AI features	50 messages every three hours for GPT-4 model	GPT-3.5, GPT-4	No	$20 per month
Kommunicate	4.50 (3.25, 5, 5, 5, 4)	E-commerce businesses	N/A	GPT-4	No	$100 per month
ChatSpot	4.40 (5, 5, 5, 5, 1)	HubSpot customers	N/A	GPT-3, GPT-4	No	Free
Intercom	4.33 (2.25, 5, 4.5, 5, 5)	Handling support queries	Unlimited – but charge per resolution	GPT-4	No	$39 per seat per month
Google Bard	4.28 (5, 5, 4.5, 5, 1)	Brainstorming ideas	Unlimited exchanges per conversation	Pathways Language Model 2 (PaLM 2)	No	Free

Name	Overall rating (cost, feature set, ease of use, quality of output, support)	Best for	Query limit	Language model(s)	Chrome extension created by vendor	Starting price
Jasper	**4.23** (2.65, 5, 5, 5, 3)	Marketing and sales team	Word limit depends on the plan	GPT-3.5, GPT-4	Yes	$49 per month
Tidio	**4.23** (4.13, 3.5, 5, 5, 4)	Small and medium businesses	Word limit depends on the plan	Claude (Anthropic AI)	No	$25 per user per month
Perplexity	**4.19** (4.15, 4.4, 5, 5, 2)	Finding information on the internet	5 Copilot searches every four hours for free users	GPT-3.5, Claude 2, GPT-4,	Yes	$20 per month
LivePerson	**4.13** (0.63, 5, 5, 5, 5)	Conversation analytics	N/A	Unknown	No	Available upon request
Chatsonic	**4.03** (4.55, 4.4, 5, 2.5, 2)	Individuals in the creative industries	Word limit depends on the plan	GPT-3.5, GPT-4	Yes	$20 per month
Drift	**3.78** (1.75, 5, 4.5, 5, 2)	Businesses that rely on B2B sales and marketing	Unlimited	GPT	No	$2,500 per month, billed annually
Ada	**3.75** (0.63, 5, 5, 5, 2.5)	Customer service automation	N/A	Unknown	No	Available upon request
YouChat	**3.64** (5, 2.9, 5, 3.75, 1)	Students and researchers	Unlimited	GPT-3, GPT-4	Yes	$6.99 per month

Name	Overall rating (cost, feature set, ease of use, quality of output, support)	Best for	Query limit	Language model(s)	Chrome extension created by vendor	Starting price
HuggingChat	3.33 (5, 2.9, 4.75, 1.25, 1)	Developers	Unlimited	Llama 2	No	$9 per month
Replika	3.16 (5, 0.63, 5, 5, 1.5)	Personal use	500 messages per month, or approximately 17 messages per day	GPT-3, GPT-4	No	$19.99 per month
Bing Chat Enterprise	3.13 (3.95, 3.5, 1.75, 2.5, 4)	Organizations in the Microsoft ecosystem	30 responses per conversation	GPT-4	No	$5 per month
OpenAI Playground	3.0 (4.15, 5, 0.55, 2.5, 2)	Customizability	200 requests per day for free users	GPT-3.5, GPT-4	No	$0.0015 per 1K tokens
Socrates.ai	2.1 (no information available, 3, 3, 2.5, 2)	Internal knowledge-base management	Unlimited	N/A	No	Available upon request

Table 1.3 The best AI chatbots on the market (data from Stefanowicz, 2024)

Name of the AI chatbot	Ratings	Free plan/trial	Best for
Jasper	5	Trial	Content creation
Elomia	4.9	Trial	Mental health assistance
Mobile Monkey	4.9	Free	Social media communication
Paradox	4.9	Custom pricing for businesses	Recruiting
Lyro	4.7	Free	Customer service features
Infeedo	4.7	Contact sales for pricing	Anonymous chats for feedback
Ada	4.6	Custom pricing for businesses	Multilingual functionality
ChatGPT	4.6	Free	Overall performance
Microsoft Bing AI	4.5	Free	Online searches
Replika	4.6	Free	Being a companion to talk to
Wati	4.6	Trail	WhatsApp customer service
Drift	4.4	Free	Scheduling options
DialogFlow	4.3	Trial	Smooth connection with Google Cloud
GitHub Copilot	4	Free	Code suggestions
Google Bard	4	Free	Creative prompts
ChatSpot	3.7	Free	Creating reports & analysis
Amazon CodeWhisperer	3.5	Free	Coding autocomplete
Snapchat My AI	N/A	Free	Speaking to AI on Snapchat
MedWhat	NA	There is no information on the developer's website	Virtual medical assistance
Kasisto	N/A	Contact sales for pricing	Finance functionalities
Personal AI	N/A	Free	Adapting to your communication style
Pi	N/A	Free	Everyday conversations

Table 1.4 The synthesis based on and indicated in three rankings of the best and the most popular AI chatbots

AI chatbots as ranked by M. Rebelo (from high to low)	The quantity indicated in 3 rankings	The overall rating	AI chatbots as ranked by A. Abdullahi (from high to low)	AI chatbots as ranked by B. Stefanowicz (from high to low)
Jasper	3	(4.23 +5)/2=**4.62**	ChatGPT	Jasper
ChatGPT	3	(4.51+4.6)/2=**4.56**	ChatSpot	Ada
Perplexity	2	**4.19**	Google Bard	ChatGPT
Ada	2	(3.75+4.6)/2=**4.18**	Jasper	Microsoft Bing AI
Google Bard	3	(4.28+4)/2=**4.14**	Perplexity	Replika
Drift	2	(3.78+4.4)/2=**4.09**	Chatsonic	Drift
ChatSpot	3	(4.40 +3.7)/2=**4.05**	Drift	GitHub Copilot
Chatsonic	2	**4.03**	Ada	Google Bard
GitHub Copilot	2	**4**	YouChat	ChatSpot
Replika	2	(3.16+4.6)/2=**3.88**	HuggingChat	Amazon CodeWhisperer
Microsoft Bing AI	3	(3.13 +4.5)/2=**3.82**	Replika	Snapchat My AI
YouChat	2	**3.64**	Bing Chat Enterprise	Personal AI
HuggingChat	2	**3.33**	OpenAI Playground	Pi
Amazon Code Whisperer	2	**3.5**		
OpenAI playground	2	**3.0**		

*: the rating can be determined by one number.

Table 1.5 The top five best/the most popular AI chatbots based on the conducted analysis of 3 ratings

The rating by M. Rebelo	The rating by A. Abdullahi	The rating by B. Stefanowicz
Jasper	ChatGPT	Jasper
ChatGPT	ChatSpot	Ada
Perplexity	Google Bard	ChatGPT
Ada	Jasper	Microsoft Bing AI
Google Bard	Perplexity	Replika

1.3. The role of AI in a higher education (following in the footsteps of UNESCO publications on the principles of the ethical use of AI systems)

In higher education institutions, AI can be beneficial for both university teachers and students (UNESCO, 2024a, 2024b). In teaching, AI can help evaluate students' tasks and create learning materials, such as tests, exercises, and even brief explanations of complex topics. It allows teachers to save time and prepare lessons with more engaging materials, create interactive presentations based on available resources, add animations, sound effects, and other elements that make the presentation more interesting, and automate many routine tasks, such as checking assignments, evaluating tests, and providing feedback to students. Students use AI to quickly find the necessary information to perform tasks and prepare for classes, learn foreign languages, generate ideas for writing essays or presentations, and analyze errors in tests (Рекомендації, 2025).

The safe use of AI in higher education is considered an essential aspect of the educational process (UNESCO, 2025). In this regard, the principles of the ethical and responsible use of AI systems in higher education are based on seven ethical values for robust AI, developed in 2019 by an independent AI HLEG expert group appointed by the European Commission. These principles are as follows: (1) human agency and oversight (AI systems should empower people with information, and proper oversight mechanisms need to be ensured; for example, automated assessment systems should contain mechanisms for reviewing and approving results by teachers); 2) technical robustness and safety (AI systems must be secure and resilient; for example, training content management systems must have data backups); (3) privacy and data governance (adequate data governance mechanisms must be ensured; for example, platforms for analyzing class activities of students should use only those data that are obtained with their consent); (4) transparency (the data and AI systems should be transparent; for example, if AI is used for personalized learning, students and teachers should understand how the algorithm makes decisions, as well as be aware of its capabilities and limitations); (5) diversity, non-discrimination and fairness (unfair bias must be avoided, and AI systems should accessible to all); (6) social and environmental well-being (AI systems should benefit all people); and (7) accountability (mechanisms should be put in place to ensure responsibility for accountability of AI systems and their outcomes; for example, universities should ensure that AI systems affecting academic or administrative decisions are regularly audited (Ethics Guidelines, 2019; Рекомендації, 2025).

Bibliography

Ali, M., & Abdel-Haq, M. K. (2021). Bibliographical analysis of artificial intelligence learning in Higher Education: is the role of the human educator and educated a thing of the past? In Fostering communication and learning with underutilized technologies in higher education (pp. 36–52). IGI Global.

Bates, T., Cobo, C., Mariño, O., et al. (2020). Can artificial intelligence transform higher education? International Journal of Educational Technology in Higher Education, 17, 42. <https://doi.org/10.1186/s41239-020-00218-x>.

Chatterjee, S., & Bhattacharjee, K. K. (2020). Adoption of artificial intelligence in higher education: a quantitative analysis using structural equation modelling. Education and Information Technologies, 25, 3443–3463. <https://doi.org/10.1007/s10639-020-10159-7>.

Cope, B., Kalantzis, M., & Searsmith, D. (2021). Artificial intelligence for education: Knowledge and its assessment in AI-enabled learning ecologies. Educational Philosophy and Theory, 53(12), 1229–1245.

Dempere, J., Modugu, K., Hesham, A., et al. (2023). The impact of ChatGPT on higher education. Frontiers in Education, 8, 1206936. <https://doi.org/10.3389/feduc.2023.1206936>.

Dhawan, S., & Batra, G. (2020). Artificial intelligence in higher education: Promises, perils, and perspective. Expanding Knowledge Horizon. OJAS, 11, 11–22.

Ethics Guidelines for Trustworthy AI (2019). European Commission. <https://digital-strategy.ec.europa.eu/en/library/ethics-guidelines-trustworthy-ai>

Forero-Corba, W., & Bennasar, F. N. (2024). Techniques and applications of Machine Learning and Artificial Intelligence in education: a systematic review. RIED-Revista Iberoamericana de Educación a Distancia, 27(1).

Ghnemat, R., Shaout, A., & Abrar, M. (2022). Higher education transformation for artificial intelligence revolution: transformation framework. Int. J. Emerg. Technol. Learn., 17(19), 224–241.

Grájeda, A., Burgos, J., Córdova, P., & Sanjinés, A. (2024). Assessing student-perceived impact of using artificial intelligence tools: Construction of a synthetic index of application in higher education. Cogent Education, 11(1), 2287917.

Hinojo-Lucena, F.-J., Aznar-Díaz, I., Cáceres-Reche, M.-P., et al. (2019). Artificial intelligence in higher education: a bibliometric study on its impact in the scientific literature. Education Sciences, 9(1), 51. <https://doi.org/10.3390/educsci9010051>.

Holmes, W., Bialik, M., & Fadel, Ch. (2023) Artificial intelligence in education. In Data ethics: building trust: how digital technologies can serve humanity. (pp. 621–653). Globethics Publications.

Jafari, F., & Keykha, A. (2023). Identifying the opportunities and challenges of artificial intelligence in higher education: a qualitative study. Journal of Applied Research in Higher Education. <https://doi.org/10.1108/JARHE-09-2023-0426>.

Mishra, R. (2019, February). Usage of data analytics and artificial intelligence in ensuring quality assurance at higher education institutions. In the Amity International Conference on Artificial Intelligence (AICAI) (pp. 1022–1025). IEEE.

O'Dea, X. C., & O'Dea, M. (2023). Is artificial intelligence really the next big thing in learning and teaching in higher education? A conceptual paper. Journal of University Teaching and Learning Practice, 20(5).

Ouyang, F., Zheng, L. & Jiao, P. (2022). Artificial intelligence in online higher education: a systematic review of empirical research from 2011 to 2020. Education and Information Technologies, 27, 7893–7925. <https://doi.org/10.1007/s10639-022-10925-9>.

Pedro, F., Subosa, M., Rivas, A., & Valverde, P. (2019). Artificial intelligence in education: challenges and opportunities for sustainable development. Paris, France, UNESCO.

Pisica, A. I., Edu, T., Zaharia, R. M., & Zaharia, R. (2023). Implementing artificial intelligence in higher education: pros and cons from the perspectives of academics. societies, 13, 118. <https://doi.org/10.3390/soc13050118>.

Рекомендації щодо відповідального впровадженні та використання технологій штучного інтелекту в закладах вищої (2025). Україна. 56 с.

Rudolph, J., Tan, S., & Tan, S. (2023). ChatGPT: Bullshit spewer or the end of traditional assessments in higher education? Journal of Applied Learning and Teaching, 6(1), 342–363.

Salinas-Navarro, D. E., Vilalta-Perdomo, E., Michel-Villarreal, R., & Montesinos, L. (2024). Using generative artificial intelligence tools to explain and enhance experiential learning for authentic assessment. Education Sciences, 14(1), 83.

Schneckenleitner, P., Hofer, M., & Reuter, F. (2023). The impact of artificial intelligence on higher education: an expert study. ICERI2023 Proceedings, 1085–1090.

Segbenya, M., Bervell, B., Frimpong-Manso, E., Otoo, I. C., Andzie, T. A., & Achina, S. (2023). Artificial intelligence in higher education: modelling the antecedents of artificial intelligence usage and effects on 21st century employability skills among postgraduate students in Ghana. Computers and Education: Artificial Intelligence, 5, 100188.

Sihare, S. R. (2024). Student dropout analysis in higher education and retention by artificial intelligence and machine learning. SN Computer Science, 5(2), 202.

Slimi, Z., & Carballido, B. V. (2023). Navigating the ethical challenges of artificial intelligence in higher education: an analysis of seven global AI ethics policies. TEM Journal, 12(2).

Sullivan, M., Kelly, A., & McLaughlan, P. (2023). ChatGPT in higher education: considerations for academic integrity and student learning. Journal of Applied Learning & Teaching, 6(1), 1–10. <https://doi.org/10.37074/jalt.2023.6.1.17>.

Taneri, G. U. (2020). Artificial intelligence & higher education: towards customized teaching and learning, and skills for an AI world of work. Research & Occasional Paper Series: CSHE. 6. Center for Studies in Higher Education.

Tao, B., Díaz, V., & Guerra, Y. (2019). Artificial intelligence and education, challenges and disadvantages for the teacher. Arctic Journal, 72(12), 30–50.

UNESCO (2024a). AI competency framework for teachers. <https://www.unesco.org/en/articles/ai-competency-framework-teachers>

UNESCO (2024b). AI competency framework for students. <https://www.unesco.org/en/articles/ai-competency-framework-students>

UNESCO (2025). Artificial intelligence in education. <https://www.unesco.org/en/digital-education/artificial-intelligence>

Vinkóczi, T., Koltai, J. P., Nagy, N. G., Szabó-Szentgróti, E., & Szabó-Szentgróti, G. (2023). The sustainable contribution of artificial intelligence to higher education: results of a pilot study. Chemical Engineering Transactions, 107, 487–492.

Zawacki-Richter, O., Marín, V. I., Bond, M., et al. (2019). Systematic review of research on artificial intelligence applications in higher education—where are the educators? International Journal of Educational Technology in Higher Education, 16, 39. <https://doi.org/10.1186/s41239-019-0171-0>.

Zhai, X., Chu, X., Chai, C. S., Jong, M. S. Y., et al. (2021). A Review of Artificial Intelligence (AI) in Education from 2010 to 2020. Complexity, 2021, 1–18.

Zhao, X., Xu, J., & Cox, A. (2024). Incorporating artificial intelligence into student academic writing in higher education: the use of Wordtune by Chinese international students.

CHAPTER 2

Microsoft Copilot and Information Technology Education

Chapter 2 describes Bing Chat (Microsoft Copilot) and provides the examples of prompts that educators can use to engage students' participation in class activities, provide hands-on training, direct the process of education, draw up lesson plans, conduct communication, give feedback, assess students' work, but to a certain extent. Through natural language processing capabilities, Copilot engages in conversations, offers suggestions, and provides assistance with creating tasks and writing assignments. Moreover, it can give instant feedback and notify users of syntax errors. It can also propose alternative answers and help teachers troubleshoot common errors. However, its, by which we mean Copilot's, ability to assess students' completed tasks is much more limited than that of human educators, as it depends on pre-programmed algorithms and may not get the full context of the student's work.

2.1. Bing Chat (Microsoft Copilot) is in layman's terms

Bing Chat (Microsoft Copilot) (Figure 2.1) is a multi-modal chatbot developed by Microsoft. Bing Chat (Microsoft Copilot) can generate text messages and images from text prompts. It to employs a neural network called Turing Natural Language Generation (abbreviated as T-NLG). This T-NLG is "a 17 billion parameter language model by Microsoft that outperforms the state of the art on many downstream NLP tasks" (Rosset, 2020).

Microsoft Copilot is a GenAI chatbot. It was established by Microsoft. Let us note that Microsoft Copilot was launched in 2023. Its distinguishing trait became an inbuilt feature which combines the functioning of Microsoft

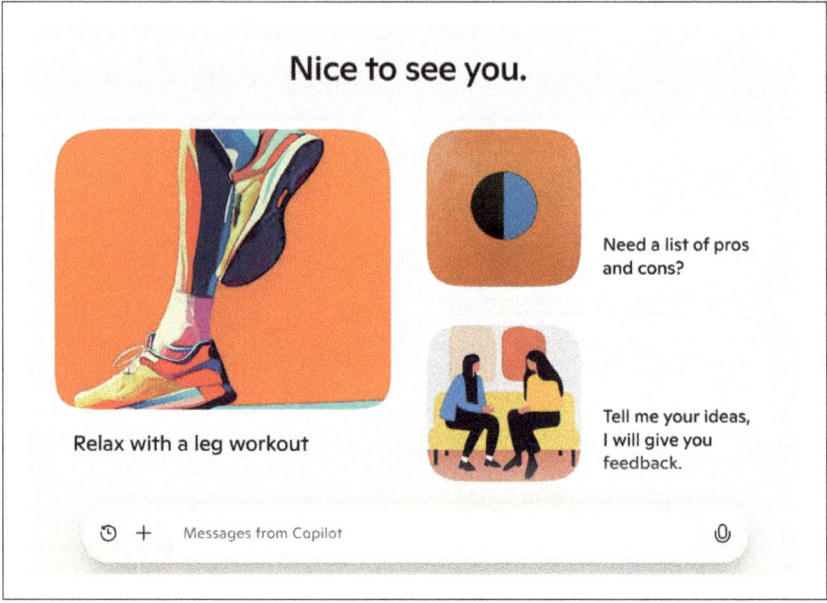

Figure 2.1 A screenshot from the website: <copilot.microsoft.com>

Copilot with Microsoft 365. "And what does it give to consumers?"—one can ask. The answer is as follows:

> Copilot works within Microsoft 365's app ecosystem which includes Microsoft Word, PowerPoint, Excel, Outlook, and Teams. As Microsoft showed us in the announcement, it can generate ideas, build reports and documents in Word and PowerPoint by pulling information from your calendar, e-mails, and contact list. In Outlook, Copilot can compose and refine emails, and in Teams, it can summarize conversations and identify tasks and assignments. Copilot can even translate live meetings and create notes and summaries. (Mauran, 2023)

It is worth noting that Microsoft 365 Copilot brings the power of AI to workplaces, as Microsoft 365 apps (i.e., Word, Excel, PowerPoint, Teams, Outlook) are used daily. Microsoft 365 Copilot is beneficial in three major directions, particularly: (1) it unleashes creativity, (2) it unlocks productivity, (3) it uplevels skills (The Official Microsoft Blog, 2023). Let us briefly characterize each of these directions. The first direction—unleash creativity—covers the creative process that Copilot in Word provides for users, particularly while Copilot gives a first draft to edit that saves hours in writing and editing time. It should be noted that sometimes Copilot's suggestions (first drafts) are helpful, but sometimes they are wrong. As an author of the content, the user must control the ideas suggested by Copilot in Word. Copilot in PowerPoint helps creating presentations with a simple prompt. Copilot in Excel analyzes trends and creates data visualization. The second direction—unlock productivity—encompasses an "office work." In particular, Copilot in Outlook can deal with inbox emails in minutes. Copilot in Teams is helpful in organizing meetings, summarizing key discussion points (including who said what, where meeting participants are aligned and where they disagreed), suggesting action items. Moreover, all these helpful actions are provided by Copilot in Teams in real time during a meeting. Copilot in PowerPoint automates repetitive tasks. The third direction—uplevel skills—allows quickly master what the user have yet to learn (a thousand of commands are available across Microsoft 365, among them "animate a slide" and "insert a table") (The Official Microsoft Blog, 2023).

On February 7, 2023, Microsoft disclosed the latest version of Bing Chat (Microsoft, 2023, February 07). Bing Chat is potentially a "game changer" that addresses some of the drawbacks and limitations of ChatGPT. The most noteworthy difference between ChatGPT and Bing Chat is that "Bing Chat is an integration of the Microsoft Bing search engine with GPT-4. The chatbot enables users to interact with the search engine using natural language. It means that instead of typing keywords and search queries, users can also ask questions or make requests in a conversational tone" (Lisowski, 2023; in Osadchiy & Osadcha, 2024). It is noteworthy that "ChatGPT is an AI-powered natural language processing model designed to generate human-like text. The model is based on the transformer architecture and was trained on massive (amount of) data from the Internet. This enables it to understand natural language, generate human-like responses, and perform various language processing tasks such as translation, summarization, question answering, and text completion" (Lisowski, 2023). Bing Chat provides footnotes with links to sources and can give proper academic references upon request. Bing's chatbot was initially in a limited preview mode while Microsoft tested it with the public, and there was a waitlist one could join for early access (Rudolph et al., 2023). It is now built into the browser Edge and easily accessible (in Osadchiy & Osadcha, 2024).

On September 21, 2023, Microsoft officially released Copilot as a follow-up to Bing Chat (Mehdi, 2023). Copilot looks at search results across the web to offer a summarized answer and links to its sources to be creative because Copilot can help write poems and stories or even create a brand-new image. With the Copilot experience, users can also ask follow-up questions such as "Can you explain that in simpler terms?" or "Can you give me more options?" to get different and even more detailed answers in search results. However, in Copilot, each conversation will have a limited number of interactions to keep the interactions grounded in search (Microsoft, 2023).

2.1.1 The descriptive characteristics of Microsoft Copilot, based on identifying and substantiating of its peculiarities

Microsoft Copilot (hereinafter for shot "Copilot") is accessible in Microsoft Edge for search of information (in Osadchiy & Osadcha, 2024). Copilot is available for the user by following the instructions, particularly:

1) click on the icon in the search field;
2) click on the chat icon in the sidebar of your browser (Figure 2.2) to get to the chat sidebar;
3) click on the arrow (Figure 2.3) to get to the full-window chat mode (Figure 2.4).

The users can switch between a full-window mode and the sidebar (Figure 2.5).

Note that to use Copilot on a mobile phone, users must download it from Google Play or the App Store (Apple).

After installing the application (hereinafter for shot "app") on the user's mobile phone, he, the user, can use Copilot's options (Figure 2.6) by clicking on the icon (Figure 2.7) of his mobile phone screen.

To use Copilot, the user needs to click in the query field (Figure 2.8).

While using the Copilot mobile app, the user can perform the same actions (explained in Osadchiy & Osadcha, 2024) as in the web app, particularly:

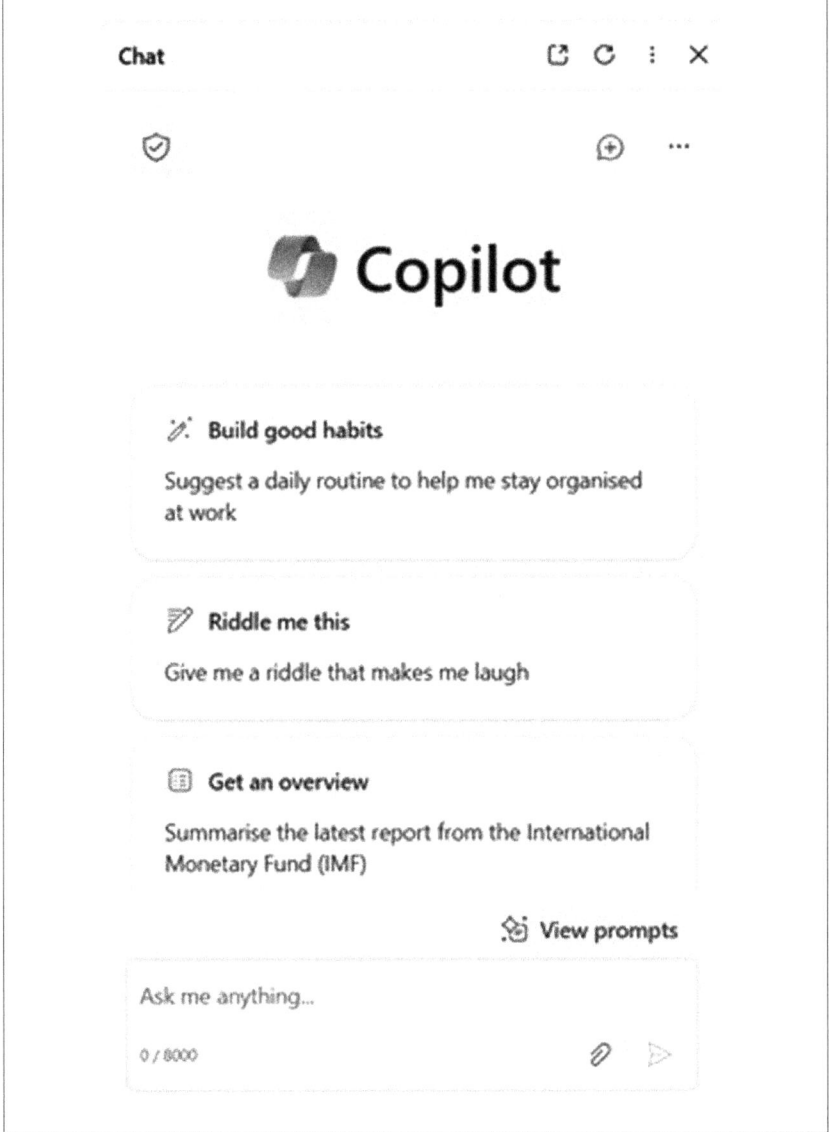

Figure 2.2 A sidebar of Microsoft Copilot's browser

CHAPTER 2

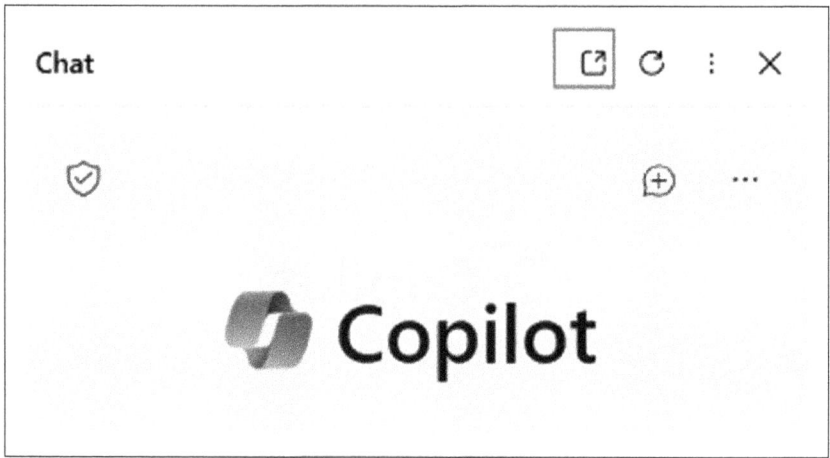

Figure 2.3 Switching to a full-window chat mode

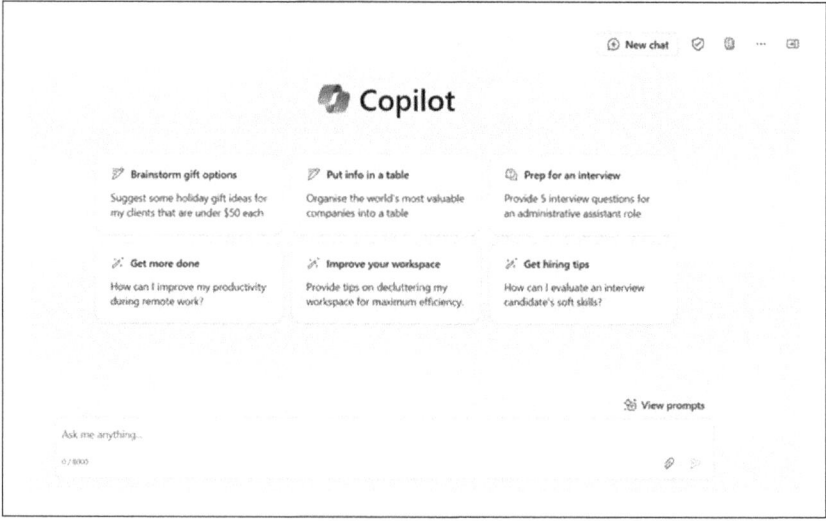

Figure 2.4 A full-window chat mode

Figure 2.5 Switching between a chat page and a traditional search page of Copilot

Figure 2.6 An icon of Copilot's app

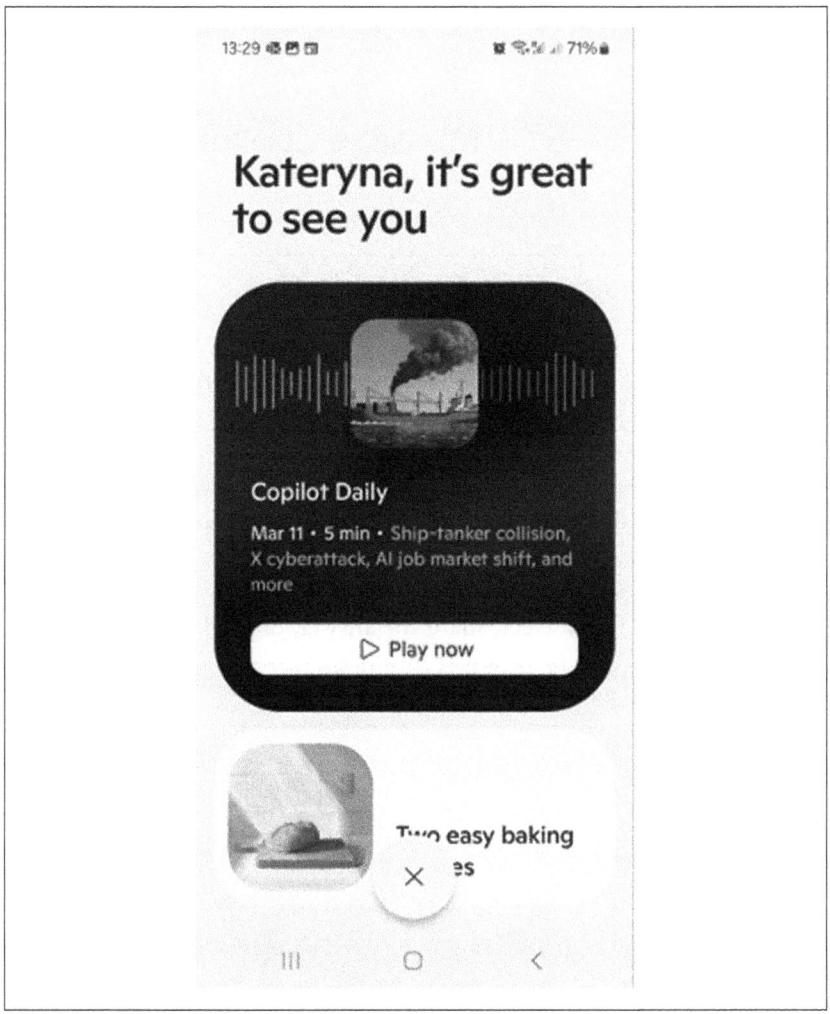

Figure 2.7 A Copilot app (Android)

Figure 2.8 A query field of Copilot's app

1. write a question and get an answer for it (Figure 2.9)
2. ask to generate an illustration, an image, and then get an image to a result (Figure 2.10)
3. the user can request image recognition; for this the user has to either take a photo of the picture (Figure 2.11) or upload an image from the mobile phone's gallery (Figure 2.12).
4. the user can use a voice input (Figure 2.13)
5. the user can copy, select, share the answer, or send feedback (Figure 2.14)

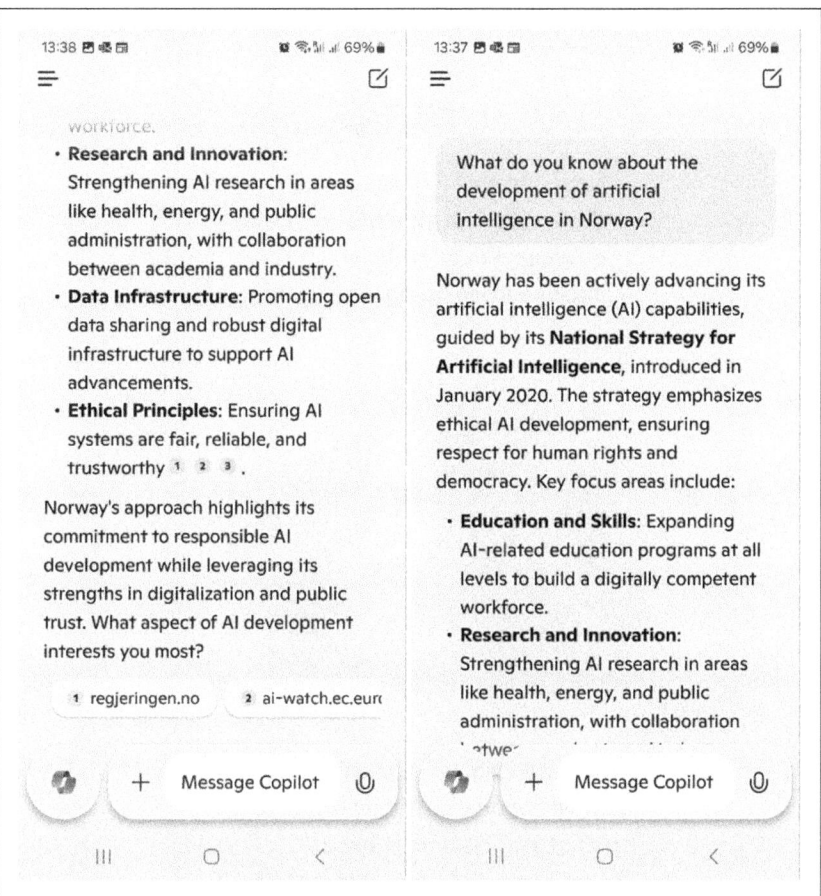

Figure 2.9 A text query and a response from Copilot

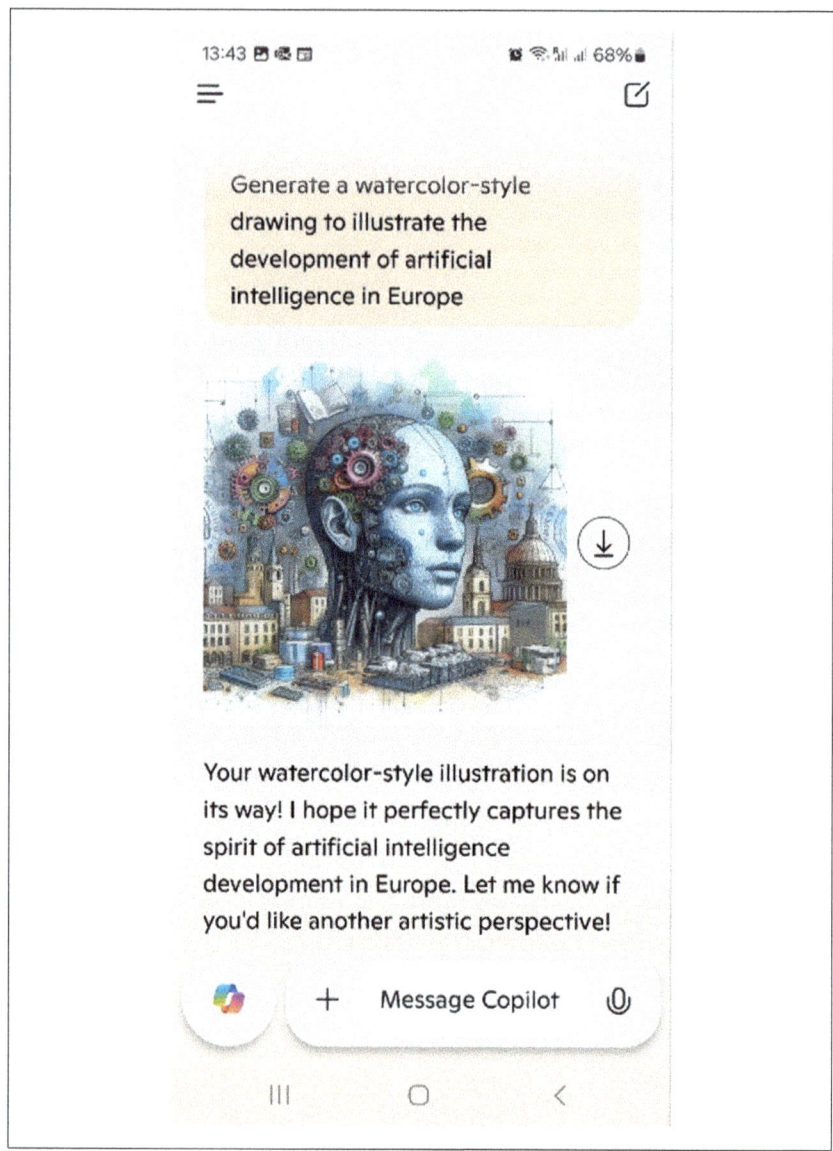

Figure 2.10 A request to Copilot for an image generation and its result

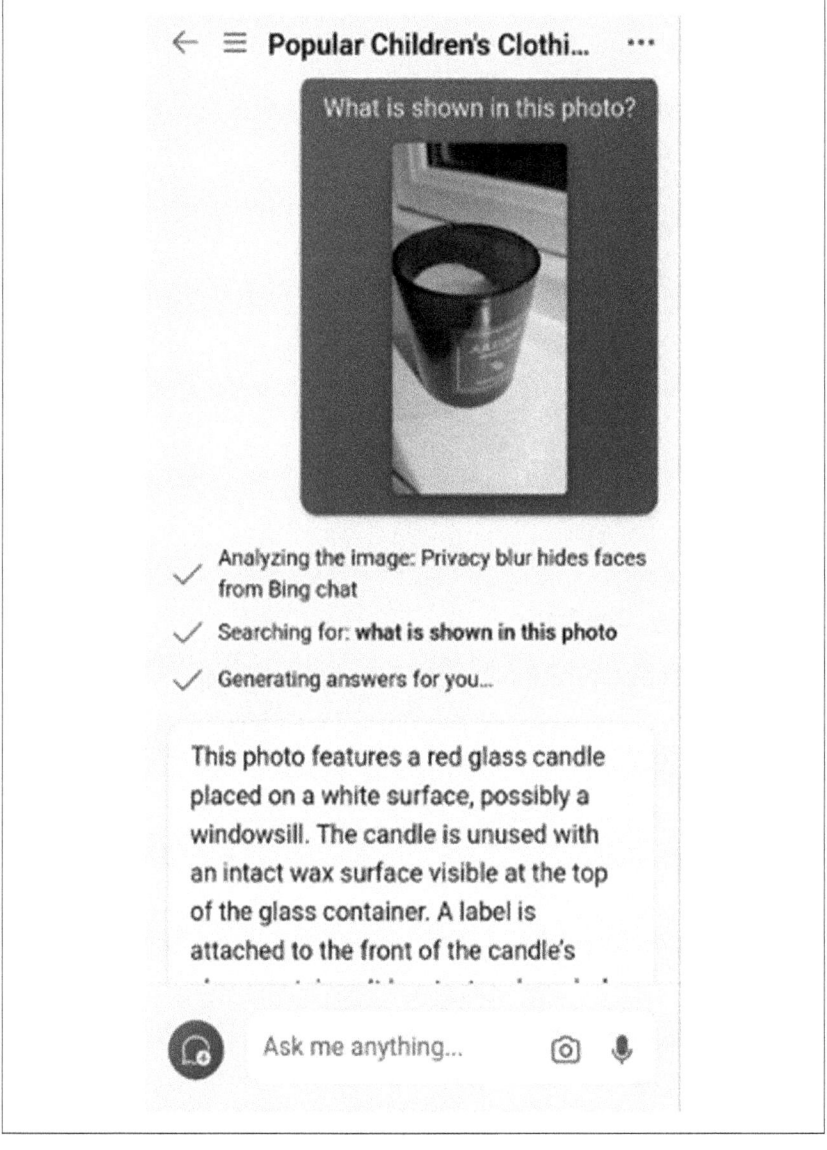

Figure 2.11 The use of an image to "communicate" with Copilot (taking a photo)

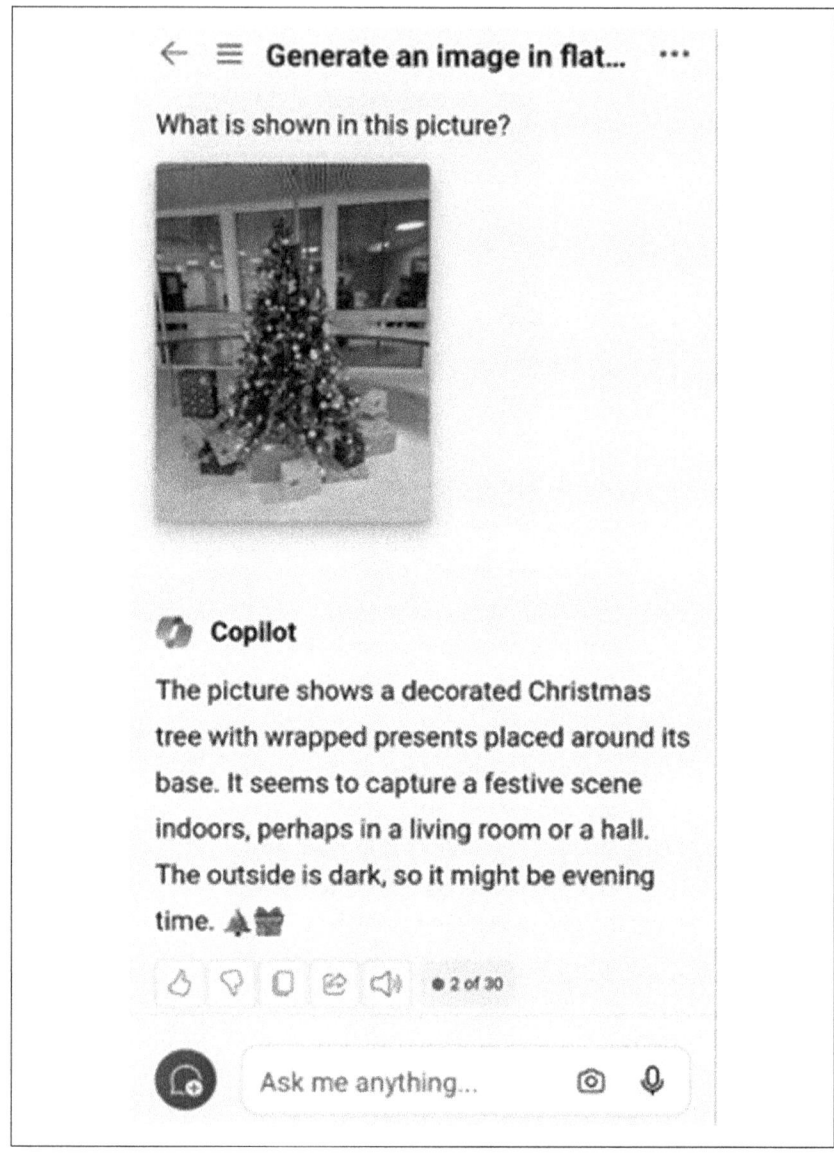

Figure 2.12 The use of an image to "communicate" with Copilot (downloading an image)

Figure 2.13 A voice input in Copilot

CHAPTER 2

Figure 2.14 The range of possible actions (i.e., copy, share, good response) in Copilot

2.1.2 How to use Copilot

When the user opens the Edge browser (<https://www.bing.com/>), in the search field, he can choose between traditional search or Microsoft Copilot Chat. After selecting and pressing the icon "Chat," the user can search for the information with the help of AI (in Osadchiy & Osadcha, 2024). The user can ask Copilot a question in one of three ways: (1) by typing the question in the prompt field, (2) by using a microphone, and (3) by using an image. To write a prompt, the user selects the prompt field (Figure 2.15), clicks on it and

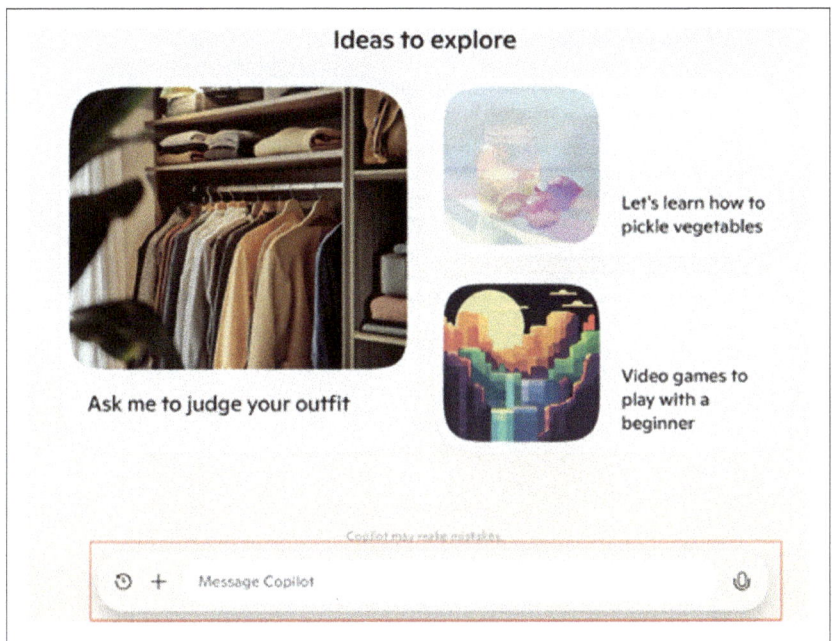

Figure 2.15 A prompt box of the MS Copilot Chat

types the questions, then Copilot can generate the answer based on the user's prompt providing a detailed information to the user's request (Licuan, 2023).

It may take up to a minute to get an answer. In addition to providing the answer to the question, the Copilot gives the sources from which the information was obtained. Also, Copilot suggests additional questions that may interest the user (Figure 2.16).

If the user doesn't want to type the questions, he can click on the microphone button (Figure 2.17) and start speaking. After that Copilot will voice the answer that it will generate based on the user's queries.

Copilot allows the user to upload images or other files, and then it can ask questions about them. To activate and then use this option, the user must click the "+" icon and then "Upload" in Copilot' input panel (Figures 2.18 and 2.19).

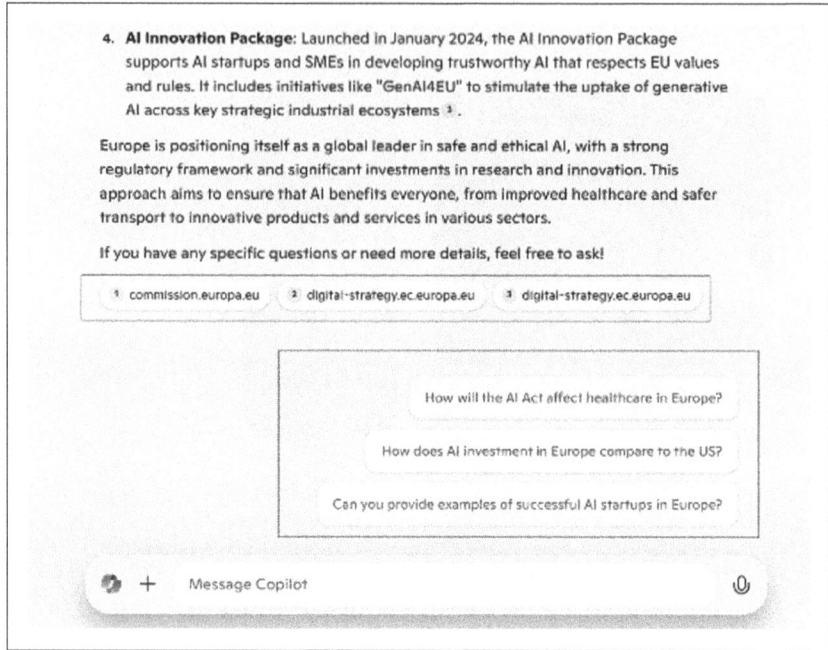

Figure 2.16 The answer of Copilot to the user's question

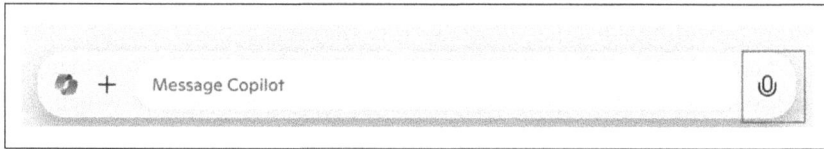

Figure 2.17 Creation of a voice request in the MS Copilot Chat

Once the answer generated by Copilot is received, the user could see the suggested follow-up questions generated by Copilot to clarify or expand the inquiry (Figure 2.20).

To obtain the answer, the user clicks on the question and get the answer (Figure 2.21).

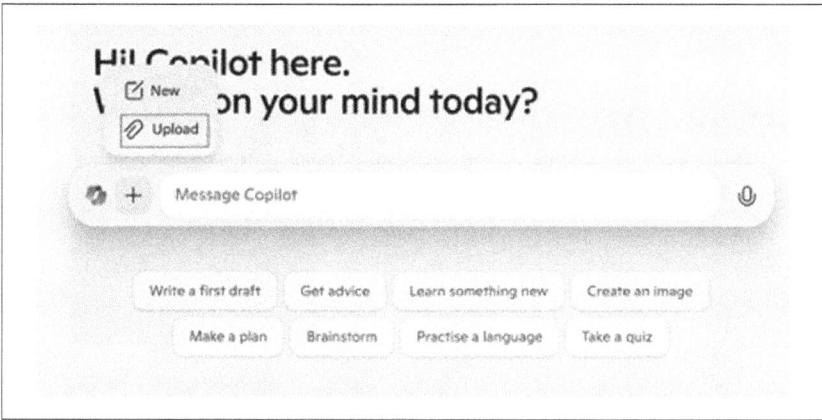

Figure 2.18 The icon in the input bar of Copilot

It is important to note that Copilot Chat can analyze the file uploaded by the user (Figure 2.22).

Moreover, Copilot can automatically generate a short description of the web page (Figure 2.23).

2.1.3 Copilot's capabilities and limitations

Copilot Chat has many capabilities, particularly it provides AI-powered suggestions, assists in content creation, and project drafting. Simultaneously, limitations in the use of Copilot should be taken into account. It is about Copilot's suggestions that should be carefully checked, and needs human oversight (including data handling). Furthermore, Copilot has (1) daily limits (Microsoft has set a limit of 30 chats per session or 300 conversations daily); (2) shallow answers (i.e., Copilot typically provides short answers to questions); and (3) ad support (Licuan, 2023).

2.2. Practically helpful prompts for Copilot

When organizing the teaching–learning process, the educator faces issues related to the organization, coordination, and conducting control over all elements of educational activity. These elements include but are not limited to planning and preparing tasks for the lesson, conducting the academic process, developing the curricula, using appropriate methods and approaches,

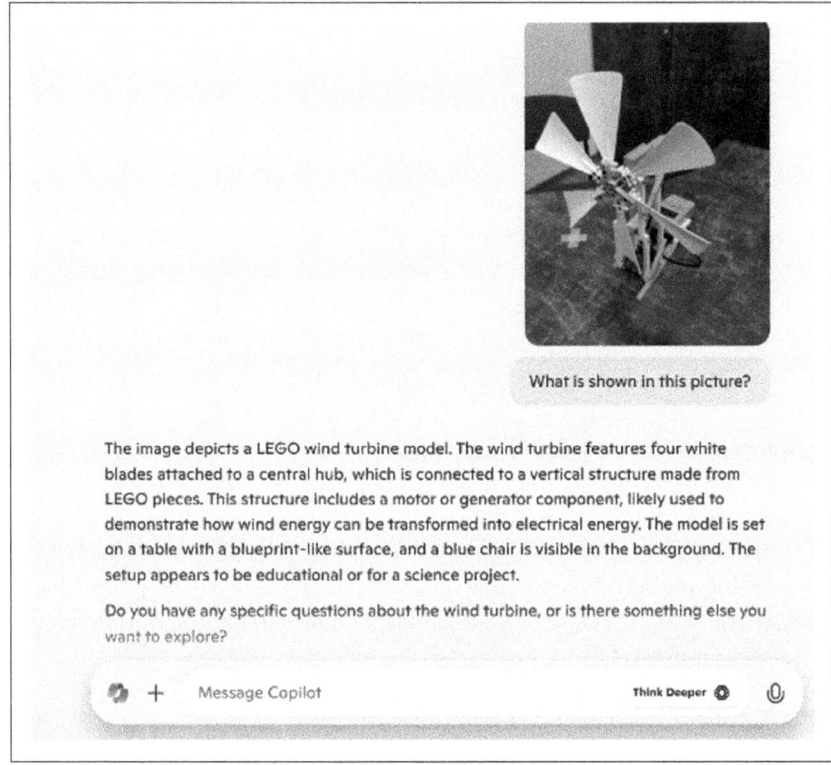

Figure 2.19 The answer of Copilot to the user's question

and writing instructional papers. All these elements are a part of education. The process of education turns out to develop a very peculiar "product" that looks different from the products in other spheres of daily activities. The product of education is a well-educated person. That is why many peculiarities characterize the process of a person's progress (in terms of their development as an individual), and the function of learning management in this context has a particular specificity. The primary objectives that the teachers set for succeeding in the development of the person are the following: goal-setting, planning of the educational process, the resource support of the educational process, ensuring high motivation of participants of the educational process,

COPILOT AND EDUCATION

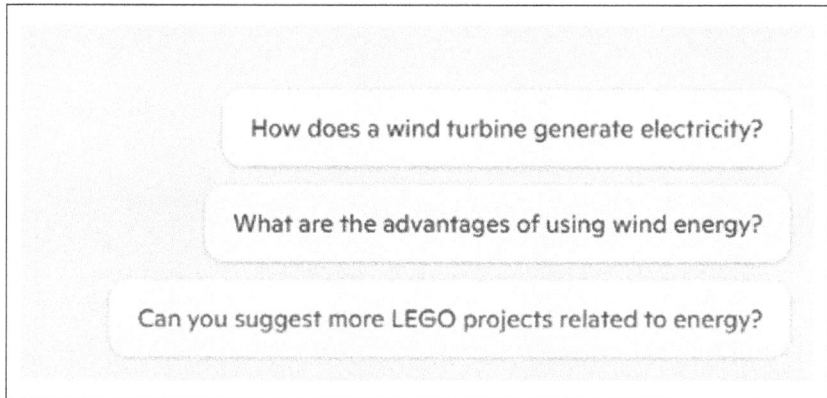

Figure 2.20 The follow-up questions generated by Copilot

> **How does a wind turbine generate electricity?**
>
> Wind turbines harness the kinetic energy from the wind and convert it into electrical energy through a series of steps:
>
> 1. **Wind blows onto the blades**: The wind turbine's large blades catch the wind's kinetic energy and begin to rotate.
> 2. **Rotor spins the shaft**: The blades are connected to a central hub, known as the rotor. As the rotor spins, it turns a shaft located inside the turbine.
> 3. **Gearbox increases rotation speed**: The spinning shaft is connected to a gearbox, which increases the rotation speed to a level suitable for generating electricity. For example, it might convert the slow rotation of the blades (around 20 rpm) into a much faster rotation for the generator (around 1,500 rpm).
> 4. **Generator produces electricity**: The high-speed rotation from the gearbox turns the generator. Inside the generator, magnets and coils of wire work together to convert mechanical energy into electrical energy through electromagnetic induction.
> 5. **Electrical energy is transmitted**: The generated electrical energy is sent through power lines to homes, businesses, and other facilities where it can be used.
> 6. **Control systems**: Modern wind turbines have control systems that adjust the angle of the blades and the direction of the turbine to optimize the amount of wind energy captured.

Figure 2.21 The response of Copilot to one of the follow-up questions

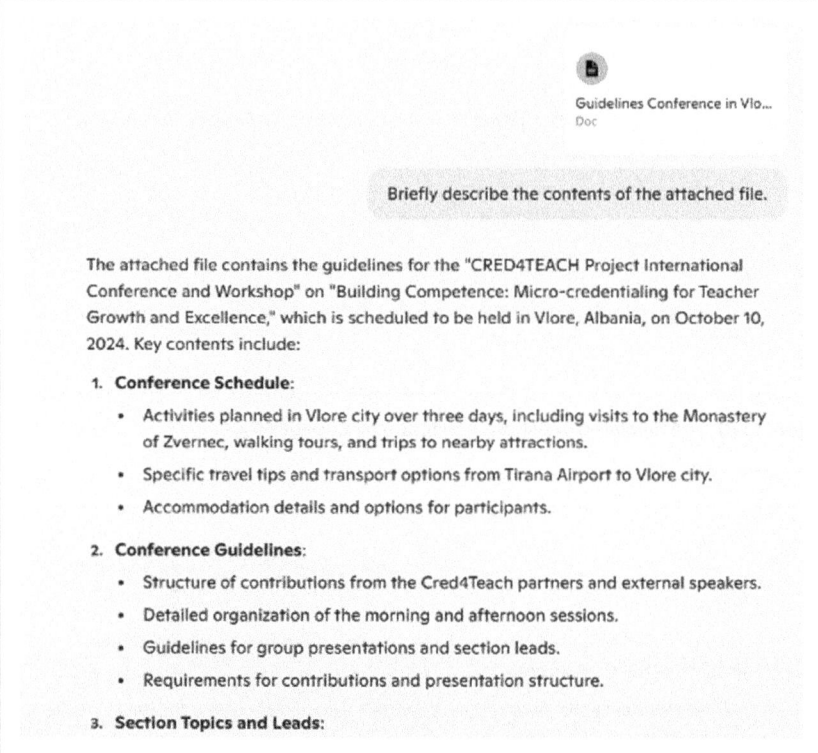

Figure 2.22 The Copilot's analysis of the file (Guidelines Conference. doc) uploaded by the user

control of academic activity, coordination of the learning process, and performance analysis (Vynnychuk, 2020).

Copilot provides valuable information that is helpful for educators, and is a source of ideas. It also helps solve specific issues. In this chapter the examples of prompts are provided for dealing with the following tasks: writing a speech, improving grammar, creating a route, drawing the plan of the route, writing lesson plans, creating a comparative table, creating a set of flesh cards, conducting assessment, making quizzes, developing writing skills and improving professional level, planning and organizing workshops, deciding research tasks, creating images and vector illustrations, making comics and logos, completing dialogues, writing poems.

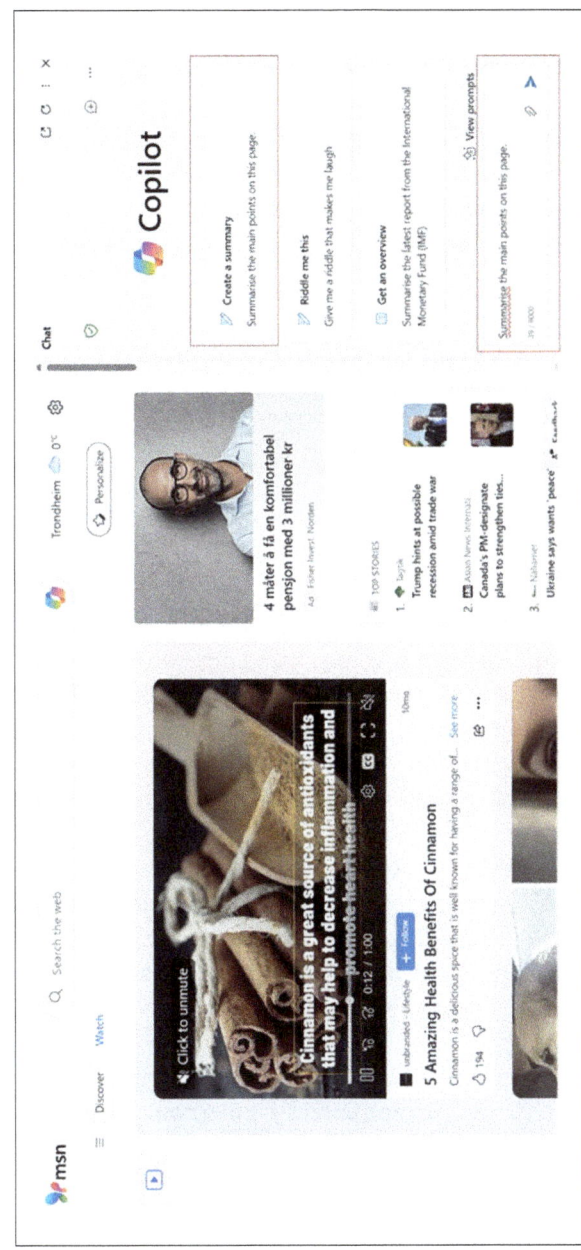

Figure 2.23 Creating a summary of the web page

Let us give examples of prompts that the teachers can use to organize the process of education effectively at the tertiary educational level.

Writing a speech

Let's take a closer look at an example that is more relevant to teaching practice: creating a text to greet students at the first class.

Prompt:
Imagine that you are a teacher, and tomorrow you need to meet with first-year students. You need to give a motivational speech to students who will study the subject "Introduction to Programming."

Result:
Absolutely! Here's my motivational speech to first-year students:

Good morning, future programmers!
Welcome to the world of programming, where your creativity and curiosity will shape the future. As you embark on this journey, remember that programming is not just about writing code; it's about solving problems, thinking critically, and turning your ideas into reality.
Programming is like learning a new language, but instead of using words to communicate with people, you'll use code to communicate with machines. It might seem challenging at first, but with persistence and practice, you'll soon find yourself speaking this new language fluently.
You'll encounter obstacles and bugs along the way, but each challenge is an opportunity to grow. Embrace those moments of frustration, for they will lead to moments of triumph when you finally crack the code.

Remember, no great programmer started as an expert. They all began where you are now, with a desire to learn and a passion for innovation. Be patient with yourself, and don't be afraid to make mistakes – they are an essential part of the learning process.
Collaboration is key in this field. Don't hesitate to ask for help, share your ideas, and learn from your peers. Together, you can achieve incredible things.
Lastly, never lose sight of the bigger picture. The skills you acquire here can change the world. From developing life-saving medical software to creating fun and engaging games, the possibilities are endless. So, dive in with enthusiasm, stay curious, and most importantly, have fun! The world of programming awaits you, and I can't wait to see the amazing things you will create.
Let's start this incredible journey together!

I hope this speech helps to inspire your students and set the stage for an exciting semester ahead! If you need anything else, just let me know.

Improving grammar

When writing different texts, readability comes with grammatically correct content (AJE's, 2025; DuBay, 2004; ProWritingAid, 2025). To provide empirical evidence supporting the claim that correct grammar enhances readability, we refer to AJE Grammar Check Tool (AJE's, 2025). This tool compares academic papers for clarity, readability, grammar, and consistency. It highlights how proper grammar contributes to better understanding and readability of research papers. It is important to note that the document entitled "The Principles of Readability" (DuBay, 2004) discusses various readability studies and emphasizes the importance of correct grammar, punctuation, and spelling in enhancing readability. Moreover, it is notable that "ProWritingAid Readability Checker" (2025) is a tool uses various readability tests and identifiers grammar issues, offering suggestions to improve readability. These sources provide empirical data and tools that demonstrate the positive impact of correct grammar on readability.

Copilot, along with the above resources (AJE's, 2025; ProWritingAid, 2025), is one of the resources that can help improve grammar. For example, enter "Improve the grammar" and add your text. Now, it will review the text, check for grammar errors, and fix them. You can also ask Copilot to rephrase the text and make it formal, professional, casual, funny, etc. Overall, Microsoft's AI chatbot can generate and enhance the text.

Prompt:
Improve the grammar of this text. The tech community, scientists, educators, and many people in general were impressed by the release of the ChatGPT neutral network. It can communicate with people very naturally, write books, theses, and articles at the level of copywriters, and create simple but working code. After that, other developments of well-known and unknown companies in the field of artificial intelligence (AI) became available: Google AI is a one-stop shop for everything related to artificial intelligence; Bing AI is a language model and graphics generation tool; DALL-E 2 is a graphics generation tool; Tome is a website that uses AI to visualize ideas and create presentations; Beatoven.ai uses advanced AI generation methods to create unique music based on mood, etc.

Result:
Sure thing! Here's an improved version of the text:

The tech community, scientists, educators, and many people in general were impressed by the release of the

ChatGPT neural network. It can communicate with people very naturally, write books, theses, and articles at the level of professional copywriters, and create simple but functional code. Following this, other developments from well-known and emerging companies in the field of artificial intelligence (AI) became available: Google AI is a one-stop shop for everything related to artificial intelligence; Bing AI is a language model and graphics creation tool; DALL-E 2 is a graphics generation tool; Tome is a website that uses AI to visualize ideas and create presentations; and Beatoven.ai uses advanced AI generation methods to create unique music based on mood.

Hope that helps! Let me know if you need anything else.

Creating a route

Sometimes teachers and scientists become conference organizers. Copilot can help create a travel itinerary.

Prompt:
Create a 1-day travel itinerary to Trondheim, it should be affordable, easy to walk, and have good public transportation.

Result:
Here's a detailed, affordable, and walkable 1-day itinerary for Trondheim that makes the most of the city's charm and its reliable public transportation system. This plan is designed to immerse you in Trondheim's history, vibe, and local delights—all while keeping things budget-friendly.

CHAPTER 2

Morning: 9:00 AM – 12:00 PM

Time	Activity	Notes
9:00 AM	Breakfast Kickoff	Start your day at a local café such as *Godt Brød* or a nearby bakery. Enjoy a light Norwegian pastry and a fresh coffee to fuel up.
10:00 AM	Nidaros Cathedral (Nidarosdomen)	Walk (or take a short bus ride from the city centre) to this Gothic masterpiece. Admire the intricate details and soak up centuries of history.
11:00 AM	Stroll to Gamle Bybro (Old Town Bridge)	Enjoy a scenic, leisurely walk to this iconic wooden bridge, offering postcard views of the Nidelva River and the charming cityscape.
11:30 AM	Explore Bakklandet	Wander through this quaint neighbourhood with colorful wooden houses, narrow alleys, and inviting coffee spots. Perfect for a photo break!

Midday: 12:00 PM – 2:00 PM

Time	Activity	Notes
12:00 PM	Lunch in Bakklandet	Savor a tasty, budget-friendly meal at one of the cosy cafés. Order a traditional Norwegian open-faced sandwich (smørbrød) or a fresh salad.
1:00 PM	Rockheim – The National Museum of Popular Music	Hop on a local bus or catch a quick tram ride to Rockheim to explore interactive exhibits showcasing Norway's vibrant music scene.

Tip: If you prefer art over music, consider a stroll to a small local gallery instead!

Afternoon: 2:00 PM – 5:00 PM

Time	Activity	Notes
2:00 PM	Return to City Centre via Public Transportation	Use the efficient "AtB" buses or trams. Grab a travel pass if you plan on using transit later in the day to save money.
2:30 PM	Trondheim Torg & Erkebispegarden (Archbishop's Palace Ruins)	Discover stories of local history at the town square and nearby ruins. Look for informational plaques explaining the city's past.
3:30 PM	Coffee Break with a View	Find a café along the riverbank or in the city centre. Enjoy a slow coffee and perhaps try a local pastry while people-watching.

Insider Note: Trondheim is compact enough for a relaxed walk. Let your curiosity guide you along vibrant streets filled with street art and local boutiques.

Evening: 5:00 PM – 9:00 PM

Time	Activity	Notes
5:00 PM	Scenic Walk along the Nidelva River	Enjoy the twilight and watch the city cast shimmering reflections on the water as you stroll.
6:00 PM	Dinner in Munkegata or Back in Bakklandet	Choose from a number of affordable restaurants or bistros that offer hearty local dishes—perfect for wrapping up your day.
7:30 PM	Evening Leisure or Local Event	If available, catch some live music at a local café or explore an art installation. Alternatively, enjoy a gentle post-dinner walk along the river.

Suggestion: Before calling it a night, use one of Trondheim's night buses if you're staying away from the city centre to head back to your accommodation safely.

Public Transportation & Budget Tips
- **Public Transport:**
  - **Trams and Buses:** Trondheim's public transit network (check the "AtB" app for live schedules) is efficient and interconnected. Use it for quicker transfers between far-flung attractions.
  - **Day-Pass Option:** If you plan on hopping between sites frequently, investing in a day-pass for public transit can save you money.
- **Budget-Friendliness:**
  - **Free Attractions:** Enjoy the outdoor beauty of the river, historic neighbourhoods, and time-worn architecture—all free of charge.
    - **Local Eats:** Look for lunch specials and early-bird dinners to keep meals both delicious and affordable.

Diving Deeper

For those who love a mix of culture and spontaneity, Trondheim always has something extra:

- **Local Markets:** On certain days, you might stumble upon pop-up craft or food markets—ideal for picking up unique souvenirs.
- **History & Art Walks:** As you care about the details, ask locals about hidden murals or lesser-known historical plaques scattered around the city.

Trondheim invites you to explore not just its famous landmarks but also its soul through neighborhood corners brimming with stories. Enjoy every moment, and let the city's layers of history and modern vibes inspire you long after your day is done!

Drawing the plan of the route
We asked Copilot to draw a map of this journey.

Prompt:
Draw this route on the map for this travel.

Result:
Below is an ASCII "sketch" map that outlines the rough flow of the day's route through Trondheim. Keep in mind that this schematic is not to scale but is intended to help you visualize the journey from one stop to the next.

How to Use This Map

- **Nodes:** Each box represents a major stop on your itinerary.
- **Arrows (|/▼):** These indicate the general flow from one location to the next.
- **Transit Notes:** Where the itinerary suggests a short bus or tram ride, you can check the "AtB" app or local transit maps in Trondheim to plan the best connection.

For precise navigation—especially if you want to explore off the beaten track—consider loading up Google Maps or the local transit apps on your smartphone. This visual guide gives you a bird's-eye view of the day's journey, ensuring that you can both savor the history and enjoy the local vibe comfortably.

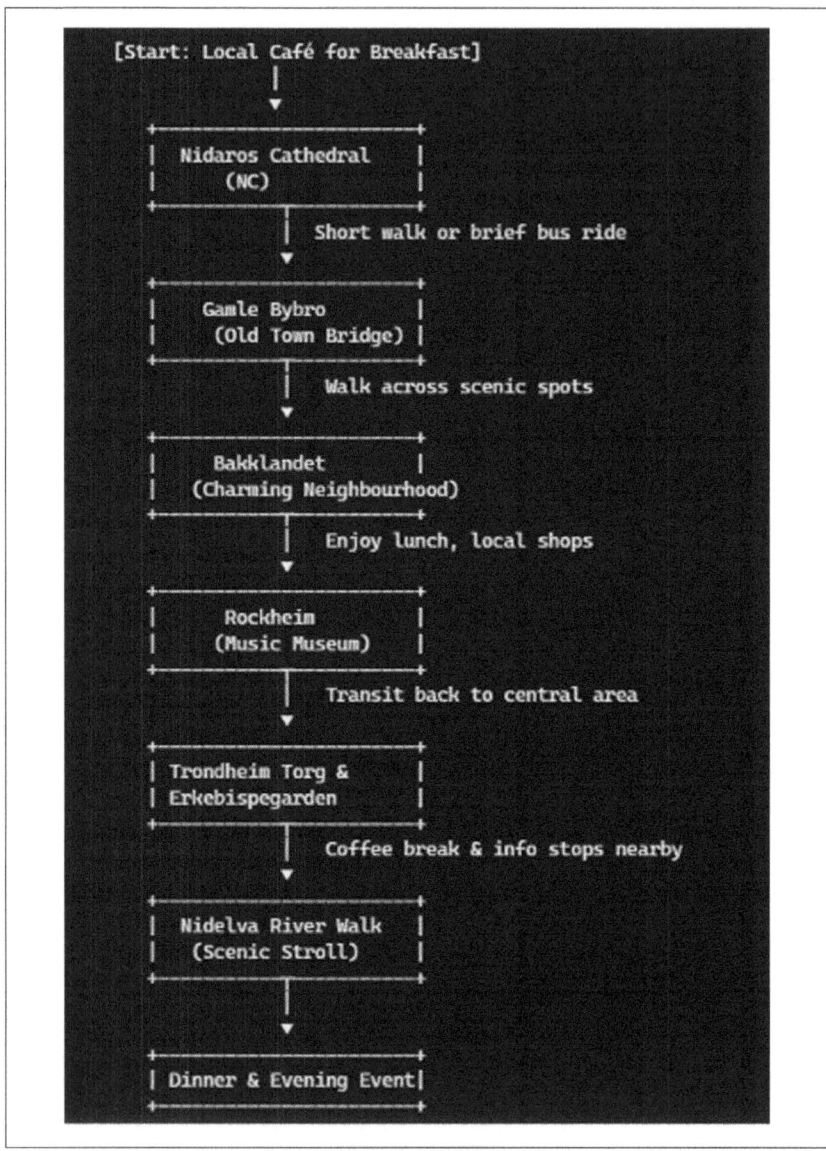

Figure 2.24 A plan of the route (compiled by Copilot)

Are you interested in a more detailed map overlay or perhaps tips on using the Trondheim transit system in real time?
Let us give one more example.

Prompt:
Plot this route on the map using https://www.bing.com/maps

Result:
Below is a pre-configured Bing Maps link that plots the major stops of your Trondheim itinerary. Simply click the link (or copy and paste it into your browser) to see the route:
View Your Trondheim Itinerary Route on Bing Maps

How to Use This Map:

1. View the Route:
   The link displays the stops in the following order:
   - **Nidaros Cathedral** (Nidarosdomen)
   - **Gamle Bybro** (Old Town Bridge)
   - **Bakklandet** (the charming neighborhood)
   - **Rockheim** (The National Museum of Popular Music)
   - **Trondheim Torg** (central town square near Erkebispegarden)

2. Choose Your Mode:
   Once the map loads, select your desired transit mode (e.g., walking, transit) by clicking the corresponding option on Bing Maps.

3. Customize Your Route:

If you'd like to add extra stops (such as your breakfast café or dinner spot) or modify the order, you can drag and drop the stops or click the "Edit" button in Bing Maps to update the route.

4. Explore and Zoom:
Feel free to zoom in on the map to check out nearby landmarks and street details. This will help you navigate Trondheim with ease.
Enjoy your journey through Trondheim—the city's rich history, scenic walks along the Nidelva River, and intimate neighborhood vibes are all at your fingertips with these directions!
Would you like more insights on local transit apps or hidden gems along the way?

Moreover, the essential tasks that require an incredible amount of effort and time in teaching activities are planning and organizing the academic process for students. To optimize these functions, teachers can use some prompts from Copilot. These prompts are well-suited to the academic teaching process.

Writing a lesson plan

Prompt:
You are a computer science educator. Come up with 5 ideas for lesson plans in our unit on the ethics of artificial intelligence. The lesson plans should be 90 minutes each and build on each other. Each lesson should be engaging and appropriate for third-year students. Each lesson should include: learning objectives; a list of key vocabulary; a lesson outline that provides for direct instruction, an engaging activity, and a closure with a synthesis/summary of student learning; and an assessment of learning.

Figure 2.25 The plan of the route (compiled by Copilot)

Result:
Absolutely, Kateryna! Here are 5 comprehensive lesson plans on the ethics of artificial intelligence for your third-year students. Each lesson is designed to be engaging and build upon the previous lesson. Let's dive in!

Lesson 1: Introduction to AI and Ethical Considerations

Learning Objectives:
- Understand the basic concepts of AI.
- Identify key ethical concerns in AI development and deployment.
- Discuss the importance of ethical AI.

Key Vocabulary:
- Artificial Intelligence (AI)
- Ethics
- Algorithm
- Bias
- Machine Learning

Lesson Outline:
1. **Direct Instruction (30 mins):**
   - Define AI and its applications in daily life.
   - Discuss historical context and development of AI.
   - Introduce ethical concerns in AI (e.g., bias, privacy, accountability).

2. **Engaging Activity (45 mins):**
   - Break students into small groups. Each group receives a case study on AI ethics (e.g., facial recognition bias, AI in hiring).
   - Groups discuss and identify ethical concerns in their case study.

- Groups present their findings to the class.

3. **Closure (15 mins):**
   - Summarize key points from group presentations.
   - Discuss the importance of ethical considerations in AI.
   - Homework: Reflective journal entry on an ethical AI issue of their choice.

Assessment:
- Group presentation rubric.
- Reflective journal entry assessment.

Lesson 2: AI Bias and Fairness

Learning Objectives:
- Understand the concept of bias in AI systems.
- Identify examples of bias in AI.
- Discuss strategies to mitigate bias in AI.

Key Vocabulary:
- Bias
- Fairness
- Discrimination
- Data Set
- Algorithmic Accountability

Lesson Outline:**
1. **Direct Instruction (30 mins):**
   - Define bias and fairness in the context of AI.
   - Present examples of biased AI systems.
   - Discuss potential impacts of biased AI on society.

2. **Engaging Activity (45 mins):**
   - Interactive demonstration: Show a biased AI model and its results.
   - Students analyze and suggest ways to mitigate bias in the model.
   - Class discussion on their findings and proposed solutions.

3. **Closure (15 mins):**
   - Recap key concepts and student suggestions.
   - Discuss the role of developers in ensuring fairness.
   - Homework: Research a real-world example of AI bias and write a summary.

Assessment:
   - Participation in the demonstration.
   - Research summary assessment.

Lesson 3: Privacy and Surveillance in AI

Learning Objectives:
   - Understand privacy concerns related to AI.
   - Identify examples of AI in surveillance.
   - Discuss the balance between security and privacy.

Key Vocabulary:
   - Privacy
   - Surveillance
   - Data Protection
   - Consent
   - Security

Lesson Outline:
1. **Direct Instruction (30 mins):**
   - Define privacy and surveillance in the context of AI.
   - Present examples of AI used in surveillance.
   - Discuss privacy concerns and data protection laws.
2. **Engaging Activity (45 mins):**
   - Debate: Split the class into two groups. One group argues for AI surveillance for security, and the other argues for privacy protection.
   - Each group presents their arguments and counterarguments.

3. **Closure (15 mins):**
   - Summarize key points from the debate.
   - Discuss the balance between security and privacy.
   - Homework: Write a persuasive essay on their stance in the debate.

Assessment:
- Debate participation rubric.
- Persuasive essay assessment.

Lesson 4: AI and Job Displacement
Learning Objectives:
- Understand the impact of AI on the job market.
- Identify industries affected by AI-driven job displacement.
- Discuss strategies to address job displacement due to AI.

Key Vocabulary:
- Job Displacement

- Automation
- Workforce
- Retraining
- Economic Impact

Lesson Outline:

1. **Direct Instruction (30 mins):**
   - Define job displacement and automation.
   - Present examples of industries affected by AI.
   - Discuss potential economic impacts of AI-driven job displacement.

2. **Engaging Activity (45 mins):**
   - Role-play: Assign students different roles (e.g., factory worker, CEO, policymaker).
   - Students discuss the impacts of AI on their roles and propose solutions.
   - Class discussion on proposed solutions and their feasibility.

3. **Closure (15 mins):**
   - Recap key concepts and student proposals.
   - Discuss the importance of workforce retraining.
   - Homework: Research and summarize an initiative addressing AI-driven job displacement.

Assessment:
- Role-play participation rubric.
- Research summary assessment.

Lesson 5: Ethical AI Development and Future Implications

Learning Objectives:

- Understand principles of ethical AI development.
- Identify future implications of AI.
- Discuss strategies for promoting ethical AI.

Key Vocabulary:
- Ethical AI Development
- Accountability
- Transparency
- Future Implications
- AI Governance

Lesson Outline:
1. **Direct Instruction (30 mins):**

   - Define principles of ethical AI development (e.g., transparency, accountability).
   - Present examples of ethical AI initiatives.
   - Discuss future implications of AI and the importance of ethical governance.

2. **Engaging Activity (45 mins):**
   - Group project: Design an ethical AI initiative addressing a specific issue (e.g., bias, privacy).
   - Groups present their initiatives and receive feedback from peers.

3. **Closure (15 mins):**
   - - Summarize key concepts and student initiatives.
   - - Discuss the role of policymakers and developers in promoting ethical AI.
   - - Homework: Reflective journal entry on the future implications of AI.

Assessment:
- - Group project presentation rubric.
- - Reflective journal entry assessment.

I hope these lesson plans will be helpful in teaching your students about the ethics of AI! Let me know if there's anything you'd like to adjust or add!

Creation of a comparative table

Prompt:
Create a comparison table between different chats that use artificial intelligence, for example, ChatGPT, Microsoft Copilot Chat, Gemini, and Perplexity AI.

Result (in tabular format for readability):

Feature	ChatGPT (OpenAI)	Microsoft Copilot Chat	Gemini (Google)	Perplexity
Developer	OpenAI	Microsoft	Google	Perplexity AI
Primary Use	General-purpose	Productivity and assistance	General-purpose	Search and knowledge assistant
Text Generation	Yes	Yes	Yes	Yes
Code Generation	Yes	Yes	Yes	No

Feature	ChatGPT (OpenAI)	Microsoft Copilot Chat	Gemini (Google)	Perplexity
Image Generation	Yes (via DALL-E)	Yes	Yes	No
Voice Mode	Yes (via Whisper)	Yes	Yes	No
Platform Availability	Web, Mobile Apps	Web, Mobile Apps, Microsoft Office	Web, Mobile Apps	Web, Mobile Apps
Special Features	Plugins, APIs, Multimodal (GPT-4o)	Integrated with Microsoft products	Deep integration with Google ecosystem, Multimodal	Real-time web data, Citation-friendly
Ideal For	Developers, Writers, General users	Professionals, Teams, General users	Professionals, Teams, General users	Researchers, Students, Fact-based queries

Let us give another example.

Prompt:
Create a comparison table between different chats that use artificial intelligence, for example, ChatGPT, Microsoft Copilot Chat, Gemini, and Perplexity AI. Use the following parameters for this – Free plan/trial, Best for What, Language model(s), and Starting price.

Result:
Below is a comparison table outlining several popular AI chat experiences along with key parameters. Keep in mind that details may evolve over time, and some values (like query limits) are subject to change or are managed on a per-use basis. This snapshot is based on information available as of late 2023:

Chat Platform	Free Plan/Trial	Best for What	Query Limit	Language Model(s)	Starting Price
ChatGPT (OpenAI)	Yes – free version available; optional ChatGPT Plus subscription (with trial periods at times)	Creative writing, general Q&A, tutoring, brainstorming, information synthesis	Varies by usage—there's no hard cap but rate limits apply during high-demand periods	**Free:** GPT-3.5 **Plus Subscribers:** GPT-4 (with improved capabilities)	**Free;** GPT-4 option at \$20/month (subject to change)
Microsoft Copilot Chat	Available through early access or previews as part of Microsoft 365; generally bundled with subscriptions	Productivity within Office apps – content drafting, data analysis, workflow assistance	Integrated within Microsoft 365 apps; limits are typically tied to overall subscription usage (no standalone cap)	Microsoft leverages proprietary solutions heavily enhanced by OpenAI technology (akin to GPT-4 derivatives)	Typically included with **Microsoft 365** subscriptions (roughly starting around \$30/month, depending on plan)

Chat Platform	Free Plan/ Trial	Best for What	Query Limit	Language Model(s)	Starting Price
Gemini (Google)	Expected to be freely accessible via Google's ecosystem (e.g., Bard environments) with beta trials	Conversational search, integrated assistance across Google services, knowledge discovery	Subject to fair use and evolving policies; specifics are still being finalized	Google's Gemini family (e.g., Gemini 1, Gemini 1.5, etc.) – models continue to evolve	Basic access through Google services is expected to be **free**
Perplexity AI	Yes – free access available	Research-focused Q&A, fact-checking, citation-backed search, and inquiry clarification	Generous usage allowances; precise limits aren't widely publicized, they focus on keeping it accessible	Utilizes a mix of OpenAI models augmented with proprietary retrieval and enhancement techniques	**Free**

Additional Thoughts
- **ChatGPT (OpenAI):** Many choose ChatGPT for its flexibility and strong conversational context, making it ideal for creative tasks as well as technical queries. The free plan is robust, and upgrading to ChatGPT Plus provides access to advanced capabilities via GPT-4.
- **Microsoft Copilot Chat:** By being deeply integrated into the Microsoft 365 ecosystem, it caters especially well to professional productivity. Users benefit from AI assistance directly inside familiar apps like Word, Excel, and Outlook. Pricing is part of a broader subscription model, so the value often comes as part of a suite of productivity tools.
- **Gemini (Google):** Integrated within Google services (for example, Bard) and leveraging Google's massive data and search infrastructure, Gemini is likely to excel at linking conversational responses with real-time search data. Its accessibility through Google's ecosystem makes it an important tool for users who rely on Google's suite for both casual and professional purposes.
- **Perplexity AI:** Distinct for its focus on research and verified sourcing, Perplexity AI appeals to users who value fact-checking and citation in answers. Its free nature makes it particularly appealing for students and researchers.

If you have questions about how these tools might fit your specific needs or want to explore technical details (or even examples of daily usage scenarios), let me know. We can further dive into nuances like integration within workflow, specialty features, or even anticipated future improvements.

Creation of a set of flash cards

Prompt:
You are an English teacher. Create a set of vocabulary flashcards on the topic "Getting away."

Result:
Sure, here are some vocabulary flash cards related to the topic "Getting away":

Vacation
Definition: An extended period of leisure and recreation, especially one spent away from home or in traveling.
Example: I'm planning a vacation to the beach this summer.

Escape
Definition: Break free from confinement or control.
   Example: The prisoner managed to escape from the jail.

Retreat
Definition: An act of moving back or withdrawing.
   Example: Sometimes, you need a quiet retreat to rejuvenate your mind.

Journey
Definition: An act of traveling from one place to another.
   Example: She's on a journey across Europe.

Adventure
Definition: An unusual and exciting or daring experience.
   Example: Going on a safari is a real adventure.

Expedition

Definition: A journey undertaken by a group of people with a particular purpose, especially that of exploration, research, or war.
  Example: The scientist led an expedition to the Antarctic.

Getaway

Definition: A period of time spent away from home or business in travel or recreation.
  Example: He planned a weekend getaway to the mountains.

Tour

Definition: A journey for pleasure in which several different places are visited.
  Example: They went on a tour of Italy.

Travel

Definition: Make a journey, typically of some length.
  Example: I love to travel to new places.

Wanderlust

Definition: A strong desire to travel.
  Example: He always had a sense of wanderlust, dreaming of far-off places.
  Remember, the best way to learn these words is to use them in your daily conversations. Happy learning! 😊

Developing a dictionary

Prompt:

You are an English teacher. Develop a 20-word dictionary for the B1 level of English on the topic of sustainable development.

Result:

I can generate a 20-word dictionary for the B1 level of English on the topic of sustainable development. Here is an example:

- biodiversity: the variety of living things in an area
- climate change: the long-term alteration of the Earth's weather patterns
- conservation: the protection and careful use of natural resources
- deforestation: the cutting down of trees in large areas
- ecosystem: a community of living and non-living things that interact with each other
- emission: the release of gas or other substances into the air
- energy: the power or ability to do work
- environment: the natural surroundings of a place
- global: relating to the whole world
- green: friendly or beneficial to the environment
- pollution: the contamination of the air, water, or soil by harmful substances
- recycle: to use something again after changing its form or purpose
- renewable: able to be replaced or restored by natural processes
- resource: something that can be used to meet a need or achieve a goal
- solar: relating to or using the energy of the sun
- sustainability: the ability to maintain or improve the quality of life without harming the environment
- waste: unwanted or unusable material or substance
- water: a liquid that covers most of the Earth's surface and is essential for life
- wind: the movement of air caused by differences in air pressure
- zero: nothing or none

Let's make the level more difficult.

Prompt:

You are an English teacher. Develop a 20-word dictionary for the C1 level of English on the topic of sustainable development.

Result:

- biodegradable: able to be broken down naturally by bacteria or other living organisms
- carbon footprint: the amount of greenhouse gases produced by a person, activity, or product
- degradation: the process of becoming worse or lower in quality
- eco-friendly: not harmful to the environment
- emissions trading: a system that allows countries or companies to buy and sell permits to emit carbon dioxide
- environmentalism: the movement or ideology that seeks to protect and improve the environment
- greenwashing: the practice of making false or misleading claims about the environmental benefits of a product or service
- innovation: the introduction of new ideas, methods, or products
- mitigation: the action of reducing the severity or harmful effects of something
- organic: produced or grown without the use of artificial chemicals
- overpopulation: the condition of having too many people living in a certain area
- resilience: the ability to recover quickly from difficulties or changes
- stewardship: the responsible management of something entrusted to one's care

- urbanization: the process of making an area more like a city
- vegan: a person who does not eat or use any animal products
- watershed: an area of land that drains water into a river, lake, or sea
- wildlife: animals that live and grow in natural conditions
- zero waste: a philosophy that aims to eliminate or reuse all materials and avoid sending anything to landfills
- zoning: the division of an area into sections that are subject to different regulations or uses
- anthropogenic: caused or influenced by human activity

A personal tutor

Prompt:
I have an English language test at the B1 level ahead of me. Can you recommend any study strategies for me?

Result:
Sure, I can recommend some study strategies for you to prepare for the B1 English test. Here are some tips:

Understand the test format and content. Review sample questions, practice exercises, and study materials provided by reputable sources123.

Assess your current language proficiency. Take online English language proficiency tests or seek guidance from language tutors to gauge your proficiency level accurately.

Create a study plan and schedule. Set specific goals and allocate dedicated time for each section of the B1 English test. Create a realistic timeline and ensure consistency in your study routine to make steady progress2.

Utilize reliable study materials and resources. Access high-quality study materials and resources designed specifically for the B1 English test. These may include textbooks, online courses, practice tests, vocabulary lists, grammar exercises, and pronunciation resources.

Practice speaking and listening skills. Find language partners or join conversation groups to improve your fluency and pronunciation. Listen to English audio materials, such as podcasts, news broadcasts, and interviews, to enhance your listening comprehension skills.

Enhance reading and writing skills. Read a wide range of English texts, such as articles, books, newspapers, and blogs, to improve your reading skills. Pay attention to comprehension, vocabulary, and understanding the main ideas. Practice writing essays, letters, and short responses to develop your writing skills, focusing on organization, coherence, and grammar.

Time management and mock tests. Simulate the test environment by practicing with timed mock tests. This will help you become familiar with the time constraints and improve your time management skills. Review your performance, identify areas for improvement, and adjust your study plan accordingly.

Seek guidance and feedback. Consider seeking guidance from experienced language tutors or enrolling in preparatory courses specifically designed for the B1 English test.

I hope these study strategies will help you ace the B1 English test. Good luck! 😊

Source: Conversation with Bing, 16.12.2023

1. B1 Preliminary preparation | Cambridge English. <https://www.cambridgeenglish.org/exams-and-tests/preliminary/preparation/>.

2. Effective Strategies for B1 English Test ... – Home Office Tests. <https://www.englishlanguagetests.co.uk/post/how-to-prepare-for-b1-english-test-1>.
3. What is the B1 English Test? A Comprehensive Guide – Home Office Tests. <https://www.englishlanguagetests.co.uk/post/understanding-the-b1-english-test>.

Creating slides prompt

Prompt:
You are an expert teacher and instructional designer. Create the content for a slide deck on "Data Visualization with Python" for my students. The slides should be formatted with a herder and then a set of bullets for each slide. Change the variety of each slide including questions for students and activities. The last slide should include text for a formative assessment. Also, include a description of any image that should be included on the slide.

Result:
Here's a draft for a slide deck on "Data Visualization with Python". Feel free to make adjustments as needed.

Slide 1: Introduction to Data Visualization
- What is Data Visualization?
- Importance of Visualizing Data
- Overview of Data Visualization Tools

Image: A colorful, engaging image of a bar chart, line graph, and pie chart to represent various data visualization types.

Slide 2: Why Python for Data Visualization?
- Python Libraries for Data Visualization
  - Matplotlib

- Seaborn
- Plotly
- Advantages of Using Python

Image: Python logo with icons of Matplotlib, Seaborn, and Plotly.

Slide 3: Setting Up Your Environment
- Installing Required Libraries
  - pip install matplotlib
  - pip install seaborn
  - pip install plotly
- Importing Libraries in Your Script

Activity: Have students install the libraries on their devices.

Slide 4: Basic Plotting with Matplotlib
- Creating a Simple Line Plot
  - Example Code
  - Customizing the Plot (Titles, Labels)
- Displaying the Plot

Image: Screenshot of a simple line plot created with Matplotlib.

Slide 5: Enhancing Visuals with Seaborn
- Introduction to Seaborn
- Creating a Scatter Plot
  - Example Code
  - Adding Trend Lines
- Customizing the Plot (Color, Style)

Question: What are the key differences between Matplotlib and Seaborn?

Slide 6: Interactive Visualizations with Plotly
- Introduction to Plotly
- Creating an Interactive Plot
  - Example Code
    - Hover Data and Interactive Elements

Image: Screenshot of an interactive Plotly graph with hover data.

Slide 7: Practical Applications
- Real-World Examples
  - Business Intelligence
  - Scientific Research
  - Social Media Analysis
- Case Studies

Activity: Divide students into groups and have them discuss how data visualization can be applied in their fields of study.

Slide 8: Best Practices in Data Visualization
- Choosing the Right Type of Chart
- Avoiding Misleading Graphs
- Keeping Visualizations Simple and Clear

Question: What are some common mistakes to avoid in data visualization?

Slide 9: Hands-On Activity
Task: Create a Data Visualization Project

- Choose a Dataset
- Create Visualizations using Python Libraries
  - Present Findings

Activity: Students work on their projects individually or in pairs.

Slide 10: Formative Assessment

Questions:

- Explain the benefits of using Python for data visualization.
- What are the main features of Matplotlib, Seaborn, and Plotly?
  - How would you approach creating a visualization for a new dataset?

Image: A blank whiteboard or a student holding a chart to represent assessment and review.

I hope this helps! If you need any more details or additional slides, feel free to ask.

Creation of a set of flashcards (another example)

Prompt:
Create a set of flashcards for an operating system dictionary.

Result:
Here are some flashcards for key terms in an operating system dictionary:

Flashcard 1
Front: Kernel Back: The core part of an operating system, managing system resources and communication between hardware and software.

Flashcard 2
Front: Process Back: A running instance of a program, which includes the program's code and its current activity.

Flashcard 3
Front: Thread Back: The smallest unit of processing that can be scheduled by an operating system.

Flashcard 4
Front: Interrupt Back: A signal that prompts the operating system to stop its current activities and execute a specific task.

Flashcard 5
Front: Virtual Memory Back: A memory management technique that gives an application the impression it has contiguous working memory, while physically it may be fragmented.

Flashcard 6
Front: File System Back: The method and data structure an operating system uses to manage files on a disk or partition.

Flashcard 7
Front: Multitasking Back: The ability of an operating system to execute more than one task simultaneously.

Flashcard 8
Front: Deadlock Back: A situation where two or more processes are unable to proceed because each is waiting for the other to release resources.

CHAPTER 2

Flashcard 9
Front: Paging Back: A memory management scheme that eliminates the need for contiguous allocation of physical memory by breaking memory into fixed-size blocks.

Flashcard 10
Front: Scheduler Back: A component of the operating system responsible for deciding which process runs at any given time.

I hope these are helpful! Let me know if you'd like more flashcards or details on any of these terms.

Assessment

Assessment of tasks that students complete in lessons or as home assignments is a complicated, laborious, routine, and undesirable process for teachers. Using Copilot's AI Chabot capabilities to evaluate the completed students' assignments will reduce teachers' psychological burden and create a form of impartial evaluation. To optimize this task, we suggest the prompts for Copilot and then evaluate the results of Copilot's suggestions and responses.

Grading and feedback on essays

Prompt:
You are an English teacher. Do an extended assessment of the essay according to the following assignment. Overpopulation of urban areas has led to numerous problems. Identify one or two serious ones and suggest ways that governments individuals can tackle these problems. [Text essay]

Result:
I can do an extended assessment of the essay according to the following assignment: Overpopulation of urban

areas has led to numerous problems. Identify one or two serious ones and suggest ways that governments and individuals can tackle these problems. Here is my feedback:

The essay has a clear structure and follows the standard format of introduction, body paragraphs, and conclusion. The introduction provides a general background of the topic and states the main argument and the outline of the essay. The body paragraphs present two problems and two solutions, each supported by relevant examples and explanations. The conclusion summarizes the main points and restates the opinion of the writer.

The essay uses a range of vocabulary and grammatical structures appropriate for the B1 level of English. The writer demonstrates the ability to use synonyms, collocations, and modifiers to express their ideas. For example, they use "overpopulation", "population growth", and "growing society" to refer to the same phenomenon. They also use "considerable dangers", "apparent complications", and "negative aspects" to describe the problems. The writer also uses complex sentences, such as conditional, comparative, and concessive clauses, to show the relationships between the ideas. For example, they use "if", "than", and "although" to introduce these clauses.

The essay is coherent and cohesive, as the writer uses a variety of linking words and devices to connect the sentences and paragraphs. The writer uses "there are", "besides this", and "however" to introduce the problems and solutions. They also use "as a result", "consequently", and "in order to" to show the cause and effect relationships. The writer also uses "for example" and "such as" to introduce the examples. The writer also uses pronouns, such as "this", "they", and "these", to refer back to the previous nouns and avoid repetition.

The essay is relevant and responds to the task adequately. The writer identifies two serious problems of overpopulation, namely the scarcity of basic goods and the unemployment rate. They also suggest two ways that governments and individuals can tackle these problems, namely introducing tax incentives for having one child and teaching family planning. The writer also expresses their opinion clearly and consistently throughout the essay.

The essay has some minor errors and areas for improvement. The writer could use more specific and accurate words to describe the problems and solutions. For example, they could use "urbanization" instead of "the above-mentioned trend" and "overconsumption" instead of "demanding necessities". They could also use "metropolitan areas" instead of "megalopolises" and "birth control" instead of "family planning". The writer could also avoid using vague words, such as "some", "a number of", and "various", and use more precise words, such as "many", "several", and "different". The writer could also check the spelling and punctuation of some words, such as "Kyiv" and "sex education".

Overall, the essay is well-written and meets the criteria for the B1 level of English. The writer demonstrates the ability to write a clear, coherent, and relevant essay on the topic of overpopulation. The writer could improve their essay by using more specific and accurate vocabulary and by proofreading their work for minor errors. I hope this feedback will help you improve your writing skills. 😊

Quizzes

Prompt:
You are an expert educator with proficiency in creating and administering student assessments. Create a 10-question multiple-choice quiz for first-year students learning about Methods for ordering, searching, and sorting. Include key concepts, vocabulary terms, and skills to analyze the efficiency of an algorithm to achieve good solutions for a given problem. Provide an answer key for the educator.

Result:
Here's a 10-question multiple-choice quiz on Methods for Ordering, Searching, and Sorting for first-year students. I've also included an answer key at the end for your convenience.

1. What is the primary purpose of sorting algorithms?

   A) To find a specific element within a data structure
   B) To reorganize elements in a certain order
   C) To reduce the size of a data structure
   D) To increase the efficiency of data storage

2. Which of the following is NOT a comparison-based sorting algorithm?

   A) Bubble Sort
   B) Quick Sort
   C) Merge Sort
   D) Counting Sort

3. In a Binary Search, what is the time complexity in the worst-case scenario?

A) O(n)
B) O(n log n)
C) O(log n)
D) O(1)

4. Which sorting algorithm is best suited for nearly sorted data?

   A) Bubble Sort
   B) Insertion Sort
   C) Selection Sort
   D) Heap Sort

5. What is the main disadvantage of using Bubble Sort?

   A) It requires additional memory
   B) It is not a stable sort
   C) It has a high time complexity
   D) It cannot sort large data sets

6. Which of the following best describes the concept of 'Big O' notation?

   A) It measures the best-case scenario of an algorithm
   B) It measures the space efficiency of an algorithm
   C) It represents the asymptotic upper bound of an algorithm's time complexity
   D) It denotes the exact running time of an algorithm

7. Which search algorithm splits the data in half to find the target value?

   A) Linear Search
   B) Exponential Search

C) Binary Search
D) Interpolation Search

8. What does it mean for a sorting algorithm to be "in-place"?

   A) It requires additional memory to sort
   B) It can sort linked lists effectively
   C) It sorts the data without needing extra space
   D) It uses a divide and conquer approach

9. Which sorting algorithm has the best average-case time complexity?

   A) Bubble Sort
   B) Quick Sort
   C) Merge Sort
   D) Insertion Sort

10. What is the primary advantage of using a Hash Table for searching?

    A) It provides constant-time search operation
    B) It is easy to implement
    C) It requires minimal memory
    D) It supports range queries

Answer Key:
 B, D, C, B, C, C, C, C, B, A

Feel free to modify the questions or add more to tailor it to your curriculum. Let me know if there's anything else you need!

CHAPTER 2

Formative assessments

Prompt:
Generate a list of 10 formative assessment ideas for first-year students who learn the course "Introduction to Operating System." These ideas must be related to the use of well-known design methods to construct new efficient algorithms.

Result:
That sounds like a great way to engage first-year students! Here are 10 formative assessment ideas related to the use of well-known design methods in constructing efficient algorithms:

1. **Algorithm Design Challenge:** Present students with a real-world problem and have them design an algorithm using methods like divide-and-conquer, dynamic programming, or greedy algorithms. They must explain their design process and justify their choices.
2. **Peer Review Sessions:** Have students exchange their algorithm designs with classmates and provide constructive feedback based on efficiency, clarity, and design principles. This promotes collaboration and critical thinking.
3. **Algorithm Optimization Task:** Provide students with an existing algorithm and ask them to optimize it using different design methods. They should compare the original and optimized versions in terms of time and space complexity.
4. **Case Studies Analysis:** Assign students to analyze case studies of well-known algorithms (e.g., Dijkstra's, Quicksort). They must identify the design methods used and discuss their efficiency and applicability to different problems.

5. **Coding Sprints:** Organize coding sprints where students are given a series of algorithmic challenges to solve within a limited time frame. Encourage them to apply specific design methods to construct efficient solutions.
6. **Algorithm Design Reports:** Ask students to write reports on their algorithm designs, detailing the problem, design methods used, and performance analysis. Provide structured feedback to help them improve their understanding.
7. **Group Projects:** Form small groups and assign them complex problems requiring efficient algorithm design. They must collaborate, apply design methods, and present their solutions to the class.
8. **Interactive Workshops:** Conduct workshops where students can brainstorm and experiment with different design methods to solve algorithmic problems. Encourage hands-on learning and iterative improvement.
9. **Algorithm Comparison:** Provide multiple algorithm solutions to the same problem and have students compare them using criteria such as efficiency, scalability, and ease of implementation. They should discuss the strengths and weaknesses of each design method.
10. **Simulation and Testing:** Have students simulate their algorithms and test them with various input data sets. They must analyze the results, identify bottlenecks, and suggest improvements based on design principles.

These formative assessment ideas will not only help students grasp the concepts of algorithm design but also encourage them to think critically and creatively while applying their knowledge. Do any of these ideas stand out to you?

CHAPTER 2

Summative performance tasks

Prompt:
You are an expert educator. You are proficient in crafting unique educational assessments. Generate 10 creative, summative performance tasks, that are an alternative to standard test tasks, for my first-year students who are learning the course "Introduction to Operating System." These tasks should be engaging, applicable, appropriate, and contribute positively to learning outcomes. These tasks should ensure students are exhibiting and applying their understanding of data structures for efficient retrieval of data and skills of dynamic programming and using retrieval of data and skills of dynamic programming and greedy algorithms.

Result:
I would be thrilled to help you design engaging and innovative performance tasks for your students. Here are 10 creative alternatives to a standard test that can effectively assess their understanding of data structures, dynamic programming, and greedy algorithms:

1. **Code an Interactive Visualizer:** Have students develop an interactive visualizer for a data structure (e.g., trees, graphs) or algorithm (e.g., dynamic programming, greedy). The visualizer should allow users to input data and visually observe how the data structure or algorithm processes the input.
2. **Algorithm Storybook:** Students can create a digital storybook that explains the concept and application of a specific algorithm or data structure. Each "page" of the storybook should include visual representations, code snippets, and real-world examples.
3. **Hackathon Challenge:** Organize a hackathon where students must solve real-world problems using data

structures and algorithms. Provide various challenges that require the application of dynamic programming and greedy algorithms to optimize solutions.
4. **Peer Teaching Videos:** Assign students to create instructional videos where they teach a particular data structure or algorithm. They should explain its principles, demonstrate code implementation, and discuss practical applications.
5. **Optimization Project:** Have students work on a project where they must optimize a specific problem using both dynamic programming and greedy algorithms. They should compare the efficiency of each approach and present their findings in a report.
6. **Simulation Game:** Students can design and develop a simulation game that utilizes data structures and algorithms to manage game mechanics. For example, they can create a resource management game where players must use algorithms to optimize resources.
7. **Interactive Tutorials:** Ask students to develop interactive tutorials that teach others about a particular data structure or algorithm. The tutorials should include interactive code examples, quizzes, and animations to engage learners.
8. **Algorithm Case Study:** Assign students to conduct a case study on the use of a specific algorithm or data structure in a real-world application. They should research how the algorithm is used, its benefits, and any challenges faced in its implementation.
9. **Algorithm Animation:** Have students create animations or visual explanations of how an algorithm works. They can use tools like Python's Matplotlib or JavaScript libraries to create step-by-step visualizations.
10. **Collaborative Coding Project:** Form student teams to work on a collaborative coding project that

```
requires the integration of multiple data struc-
tures and algorithms. They should document their
development process and present their final product
to the class.
```

```
These tasks aim to encourage hands-on learning, crea-
tivity, and collaboration while ensuring students can
practically apply their knowledge. They also provide
a variety of ways for students to demonstrate their
understanding, catering to different learning styles.
   Would you like any further details or adjustments
on these tasks?
```

Communication functions

One of the essential functions in the teaching and learning process is the communicative function. In this context, the educational process's quality is evidenced by students' ability to speak and express complex ideas and thoughts in a foreign language. Moreover, interpersonal relationships and business ties are formed and strengthened between the participants of the communication process through effective communication. Furthermore, effective communication between the teacher and the student or between the students in the class optimizes and improves the efficiency of the educational process and contributes to the implementation of many principles of developing learning. To improve speaking skills, we suggest prompts to help teachers enhance the students' communication skills.

Developing writing skills

Email response prompt

Prompt:
```
Create a response to this email from a student about
setting up a time to discuss additional academic sup-
port for the next semester. Keep the tone professional
and friendly and write from the point of view of an
educator.
```

Result:
Dear [Student's Name],
   Thank you for reaching out to discuss additional academic support for the upcoming semester. I'm glad to see your proactive approach to your education, and I'm here to assist you in any way I can.
   Let's schedule a meeting to talk about the various support options available and develop a personalized plan that aligns with your academic goals. Please let me know your availability next week, and we can arrange a time that is convenient for both of us.
   In the meantime, if you have any specific concerns or areas you'd like to focus on, feel free to share them with me. This will help us make the most of our meeting time.
   Looking forward to our conversation and to supporting your academic journey.
   Warm regards,
   [Your Name] [Your Position]

Newsletter prompt

Prompt:
You are an expert educator and skilled writer. Write a weekly newsletter that I can send to my students on the topic of novelties in the field of artificial intelligence. Include sections for each day of the week. For each day of the week write a brief paragraph summarizing the main day news. Make the newsletter 1 page long and make the tone of the newsletter professional and friendly.

Result:

Weekly AI Insights Newsletter
Monday: AI Stocks on the Rise This week started with a notable shift in the stock market. Two prominent AI

stocks experienced significant dips, presenting a potential buying opportunity for investors. As AI continues to evolve, keeping an eye on market trends can provide valuable insights into the industry's financial health.

Tuesday: AI's Power Problem A critical discussion emerged about the energy consumption of AI technologies. As AI systems become more advanced, their power requirements are skyrocketing. This raises important questions about the sustainability of AI and the need for innovative solutions to address these challenges.

Wednesday: AI in Counter-Terrorism The United Nations Office on Drugs and Crime (UNODC) hosted a symposium exploring the role of AI in preventing and countering terrorism. Experts discussed both the risks and opportunities AI presents in this field, emphasizing the importance of responsible integration of AI technologies.

Thursday: AI in Healthcare A breakthrough in AI-powered healthcare was announced with the development of a new AI speech model that significantly reduces transcription errors in medical settings. This advancement promises to enhance the accuracy and efficiency of healthcare documentation.

Friday: AI and Robotics Researchers have made strides in developing robotic systems that can be controlled through brain signals. This innovation allows paralyzed individuals to manipulate robotic arms with their thoughts, showcasing the incredible potential of AI in improving quality of life.

Saturday: AI in Education AI's impact on education continues to grow, with new tools being developed to enhance learning experiences. From personalized learning platforms to AI-driven tutoring systems, the integration of AI in education is transforming how students learn and interact with educational content.

Sunday: AI and Ethics The ethical implications of AI remain a hot topic. Recent discussions have focused on the need for transparency and accountability in AI development. Ensuring that AI technologies are used responsibly and ethically is crucial for their long-term success and societal acceptance.

I hope you find this newsletter informative and engaging. If you have any questions or need further information, feel free to reach out!

Best regards, Kateryna

Feel free to adjust any details as needed! If you need further assistance, I'm here to help.

Reference letter prompt

Prompt:
You are an expert researcher, skillful at conveying a person's abilities and attributes. Write a reference letter for my student in Computer Science who is applying to the Norwegian University of Science and Technology. Use specific, positive, descriptive words, and be sure to highlight my student's outstanding academic achievement, exceptional personal qualities, and dedication to the field of Computer Science through her research project on a calculation of the carbon footprint in the programming process. Also note her extracurricular activity involvement in student government, and volunteering at Computer Science conferences. In your response do not make the reference letter more than one page and do not report specific grades.

Result:
  [Your Name]
  [Your Title]
  [Your Institution]
  [Your Contact Information]

CHAPTER 2

Date: [Today's Date]

Admissions Committee
Norwegian University of Science and Technology
Address

Dear Members of the Admissions Committee,

It is with great enthusiasm that I recommend [Student's Full Name] for admission to the Norwegian University of Science and Technology's Computer Science program. As [his/her/their] [professor/advisor], I have had the pleasure of observing [his/her/their] academic journey and personal growth over the past [number] years. [Student's Name] is an exceptional student whose commitment to academic excellence, personal integrity, and dedication to the field of Computer Science set [him/her/them] apart from [his/her/their] peers.

Academically, [Student's Name] has consistently demonstrated a deep understanding and passion for Computer Science. [His/Her/Their] research project on the calculation of the carbon footprint in the programming process is a testament to [his/her/their] ability to engage with complex topics and contribute meaningful insights to the field. This innovative project showcases [his/her/their] technical prowess, critical thinking skills, and commitment to addressing real-world challenges through Computer Science.

Beyond academics, [Student's Name] is a well-rounded individual who has actively participated in various extracurricular activities. [His/Her/Their] involvement in student government has honed [his/her/their] leadership skills and demonstrated [his/her/their] ability to collaborate effectively with others. Furthermore, [he/she/they] has generously volunteered at numerous Computer Science conferences, further solidifying [his/

her/their] dedication to the field and willingness to give back to the community.

[Student's Name] possesses exceptional personal qualities that make [him/her/them] a valuable addition to any academic environment. [He/She/They] is highly motivated, intellectually curious, and always eager to learn. [His/Her/Their] ability to balance rigorous academic pursuits with extracurricular commitments speaks volumes about [his/her/their] time management skills and dedication.

In conclusion, I am confident that [Student's Name] will thrive at the Norwegian University of Science and Technology and make significant contributions to the Computer Science community. [His/Her/Their] outstanding academic achievements, exceptional personal qualities, and unwavering dedication to the field make [him/her/them] an ideal candidate for your program.

Thank you for considering [Student's Name]'s application. If you require any further information, please do not hesitate to contact me.

Sincerely,

[Your Full Name]
[Your Title]
[Your Institution]
[Your Contact Information]

Professional development

List of tips

Prompt:
I'm a computer science educator and I'm new to using Microsoft Bing AI. Give me a list of important Bing AI tips to help me get more done and save time.

Result:
Certainly! As a computer science educator, here are some important **Bing AI tips** to enhance your productivity and make the most of your time:

1. **Microsoft Copilot in Education:**
   - Explore using **Microsoft Copilot** in your teaching. Learn about its basic concepts, modes, and features. Copilot can assist you in generating code, explanations, and more.
   - Use it to streamline lesson planning, code writing, and debugging tasks.
2. **Reading Progress with AI:**
   - **Reading Progress** helps improve reading fluency. It provides insights into student reading patterns and saves you time by automating assessments.
   - Encourage students to practice reading independently using this tool.
3. **Generative AI in Bing:**
   - Interact with **Microsoft Copilot** within **Bing**. Discover the capabilities of generative AI.
   - Use it to explore various topics, generate creative content, and find answers efficiently.
4. **AI-Powered Storytelling:**
   - **Reading Coach** engages students by creating AI-generated stories. These stories are moderated for safety and age-appropriateness.
   - Encourage students to create their own unique stories using AI.
5. **GitHub Prompt Library for Education:**
   - Utilize the **GitHub prompt library** containing curated Bing Chat prompts. These can help you craft engaging lessons, respond to student inquiries, and manage administrative tasks effectively.

6. **Learn Continuously:**
   o Join **Microsoft Learn for Educators** to access continually updated technical AI content relevant to your classroom.
   o Keep your curriculum aligned with industry needs and real-world job responsibilities.
7. **Responsible AI Usage:**
   o Teach students about responsible AI practices, including prompt engineering and ethical considerations.
   o Explore AI's potential while emphasizing responsible use.

Remember, AI can significantly enhance your teaching experience and empower your students. Feel free to explore these tools and adapt them to your specific needs! 🚀🤖

For more detailed information, you can refer to the Microsoft Learn Educator Center and the Empower Educators to Explore the Potential of Artificial Intelligence module. Happy teaching! 😊

List of prompts

Prompt ("More Balanced"):
I'm a computer science teacher and I'm new to using Microsoft Bing AI. Give me a list of important Bing AI prompts that will help me learn more about the practical aspects of Python programming for creating AI applications.

Result:
Certainly! As a computer science teacher, here are some important **Bing AI prompts** that can help you explore practical aspects of Python programming for creating AI applications:

1. **Summarize Python Libraries for AI:**
   - Prompt: "Summarize the key Python libraries commonly used for artificial intelligence and machine learning."
   - This will provide an overview of popular libraries like **NumPy, Pandas, TensorFlow,** and **PyTorch.**
2. **Python Code for Data Preprocessing:**
   - Prompt: "Provide Python code examples for data preprocessing tasks such as cleaning, normalization, and feature engineering."
   - Use this to learn about handling data before feeding it into AI models.
3. **Implementing Neural Networks in Python:**
   - Prompt: "Explain how to create a basic neural network using Python."
   - You'll receive guidance on building neural networks from scratch or using libraries like **Keras.**
4. **Natural Language Processing (NLP) Basics:**
   - Prompt: "Describe the fundamental concepts of NLP and how to process text data in Python."
   - Learn about tokenization, stemming, and sentiment analysis.
5. **Image Classification with Convolutional Neural Networks (CNNs):**
   - Prompt: "Walk me through the steps to build an image classifier using CNNs in Python."
   - Understand how to preprocess images and create CNN architectures.
6. **Deploying AI Models with Flask:**
   - Prompt: "Explain how to deploy a trained machine learning model using Flask."
   - Learn about creating a REST API for your AI applications.
7. **Handling Imbalanced Datasets:**
   - Prompt: "Discuss strategies for dealing with imbalanced datasets in Python."

- Explore techniques like oversampling, undersampling, and using synthetic data.
8. **Hyperparameter Tuning for AI Models:**
   - Prompt: "How can I optimize hyperparameters for my AI model in Python?"
   - Learn about techniques like grid search, random search, and Bayesian optimization.
9. **Time Series Forecasting with Python:**
   - Prompt: "Provide an example of time series forecasting using Python."
   - Understand how to use libraries like **Prophet** or **ARIMA**.
10. **Ethical Considerations in AI Development:**
    - Prompt: "Discuss ethical challenges related to AI development and how to address them."
    - Learn about bias, fairness, and transparency.

Remember to adapt these prompts to your specific learning goals and explore further based on the responses you receive. Happy learning! ☐☐

Planning and organizing workshops

Prompt:
You are an expert educator and supervisor, proficient in planning and organizing workshops. Generate 5 ideas for professional development presentations on the topic of education technology in IT education. The presentations should be 1 hour each. The presentations should focus on digital literacy, technological tools for teachers and how to integrate technology AI into the classroom. The presentations should be engaging, and applicable for novice educators. Each presentation should utilize principles of adult learning theory.

Result:
Certainly! Here are five engaging one-hour professional development presentation ideas for novice educators, focusing on education technology in IT education:

1. **"Navigating the Digital Landscape: Building Digital Literacy Skills"**
   - Explore the concept of digital literacy and its importance in the modern classroom.
   - Discuss strategies for evaluating digital resources for credibility and relevance.
   - Engage participants in hands-on activities using digital tools to enhance their own literacy.
2. **"Tech Toolbox: Essential Technological Tools for Today's Educators"**
   - Introduce a variety of technological tools that can aid in lesson planning, student assessment, and classroom management.
   - Provide demonstrations of select tools and discuss practical applications in IT education.
   - Facilitate a collaborative session where educators share experiences and best practices.
3. **"AI in Education: Transforming the Classroom with Smart Technology"**
   - Offer an overview of AI technologies currently available for educational purposes.
   - Illustrate how AI can personalize learning experiences and provide adaptive learning paths for students.
   - Conduct a group brainstorming session to generate ideas for AI integration in their teaching practices.
4. **"From Consumers to Creators: Fostering Digital Creation in the Classroom"**
   - Emphasize the shift from consuming digital content to creating it as a form of learning.

- Showcase platforms and tools that enable students to create digital projects, such as coding, multimedia production, and web design.
- Organize a mini-workshop where educators can experience these tools firsthand.
5. "The Connected Classroom: Strategies for Integrating Technology in Teaching"
   - Discuss the pedagogical benefits of integrating technology into the curriculum.
   - Present case studies of successful technology integration in IT education.
   - Facilitate a planning session where educators outline a technology integration plan for a unit or lesson.

Each presentation will incorporate adult learning theory principles, such as active learning, collaborative discovery, and practical application, ensuring that the content is relevant and immediately applicable for the educators attending.

Research tasks
Creating a list of references

Prompt:
I have a bibliography in APA citation style. Make it IEEE citation style.

1. *Becker, B. A., Denny, P., Finnie-Ansley, J., Luxton-Reilly, A., Prather, J., & Santos, E.A. (2023). Programming Is Hard – Or at Least It Used to Be: Educational Opportunities and Challenges of AI Code Generation. Proceedings of the 54th ACM Technical Symposium on Computer Science Education V. 1. Association for Computing Machinery, New York, NY, USA, 500–506. <https://doi .org/10.1145/3545945.3569759>.*

2. Santos, V., Mamede, H., Silveira, C., & Reis, L. (2023). *A Reference Model for Artificial Intelligence Techniques in Stimulating Reasoning, and Cognitive and Motor Development*. Procedia Computer Science, 219, 1057–1066. <https://doi.org/10.1016/j.procs.2023.01.384>.
3. Li, P. (2022). *Teaching Transformation of Computer Application Technology Specialty Based on Artificial Intelligence*. In: Hung, J.C., Chang, JW., Pei, Y., Wu, WC. (eds) *Innovative Computing. Lecture Notes in Electrical Engineering*, vol 791. Springer, Singapore.

Result:

> Got it! Here are your references in IEEE citation style:
>
> 1. B. A. Becker, P. Denny, J. Finnie-Ansley, A. Luxton-Reilly, J. Prather, and E.A. Santos, "Programming Is Hard – Or at Least It Used to Be: Educational Opportunities and Challenges of AI Code Generation," in Proc. 54th ACM Technical Symp. on Computer Science Education V. 1, New York, NY, USA, 2023, pp. 500–506. doi: 10.1145/3545945.3569759.
>
> 2. V. Santos, H. Mamede, C. Silveira, and L. Reis, "A Reference Model for Artificial Intelligence Techniques in Stimulating Reasoning, and Cognitive and Motor Development," Procedia Computer Science, vol. 219, pp. 1057–1066, 2023. doi: 10.1016/j.procs.2023.01.384.
>
> 3. P. Li, "Teaching Transformation of Computer Application Technology Specialty Based on Artificial Intelligence," in Innovative Computing, J.C. Hung, J.W. Chang, Y. Pei, and W.C. Wu, Eds., Lecture Notes in Electrical Engineering, vol. 791, Singapore: Springer, 2022.
>
> Feel free to let me know if there's anything else you need help with!

Figure 2.26 A list of references (compiled by Copilot)

List of 10 research papers

Prompt:
Make a list of 10 research papers that study the use of Microsoft Copilot Chat education.

Result:
Sure, here are some research papers that explore the use of Microsoft Copilot Chat in education:
1. [Using Microsoft Copilot Chat in the Work of IT Educators: Pilot Study](<https://ntnuopen.ntnu.no/ntnu-xmlui/handle/11250/3182235>)
2. [Microsoft Copilot and Anthropic Claude AI in education and library service](<https://www.emerald.com/insight/content/doi/10.1108/lhtn-01-2024-0002/full/html>)
3. [Using Microsoft Copilot Chat in the Work of IT Educators: Pilot Study](<https://ntnuopen.ntnu.no/ntnu-xmlui/bitstream/handle/11250/3182235/6202-Article%2bText-23836-1-10-20241124%2b%25281%2529.pdf?sequence=2>)

I could only find three relevant papers. If you need more information or specific details from these papers, feel free to ask!

An explication in simple words

Prompt:
Explain in simple words the point of this article <https://www.mdpi.com/2071-1050/16/17/7695>.

Result:
Below is a plain-language summary of the article's main idea. (Keep in mind that without reading every

detail of the full text, this explanation is based on the abstract and overall framing provided on the article's webpage.)
In simple words, the article is about:

- **Exploring a New Approach for Better Outcomes:** The study investigates a method (or strategy) that aims to improve how some organizations—often universities or community institutions—operate so that they become more supportive, caring, and sustainable over the long term.
- **Balancing High-Tech or Innovative Tools With Real-Human Needs:** While modern tools (like digital technologies or even artificial intelligence) and innovative strategies can make systems work more efficiently or help solve some problems, the article shows that these solutions don't automatically satisfy all human or social needs. For example, it may reveal that even with a new approach in place, a noticeable share of users or students still feels that something essential is missing—such as a warm, inclusive, and respectful environment.
- **Identifying Strengths and Areas for Improvement:** The research isn't just about praising a new method; it critically asks, "What works well and what falls short?" It provides evidence (from surveys, case studies, or data analysis) indicating that while the new approach has many benefits, there are also challenges or gaps that must be addressed. In short, the study encourages decision-makers (like educators, managers, or policymakers) to look beyond the "tech fix" and pay close attention to the quality of the human experience.
- **A Call for Holistic, Human-Centered Sustainability:** Ultimately, the article argues that long-term sustainability isn't just about using cutting-edge tools

or clever strategies. It also involves creating environments—whether in schools, companies, or communities—that are "non-violent" in the sense that they actively support people's well-being, respect, and personal development. In other words, true sustainability needs a balance between innovation and human care.

Why does this matter? If you're involved in shaping environments where people learn, work, or live, the article suggests that you shouldn't rely solely on technology or rigid strategies. Instead, you need to consider the full picture—including emotional, social, and ethical dimensions—to build systems that truly serve everyone over time.

This summary provides a broad overview. There's more nuance in the full article—such as the specific contexts studied, the data presented, and the detailed recommendations—but at its heart, the work is about finding that delicate balance between innovation and humanity in the pursuit of sustainable practices.

Playing games
Tic-tac-toe

Prompt ("More Precise"): `Let's play tic-tac-toe`

Result:

Trivia

Prompt ("More Precise"): `Let's play trivia.`

Result:

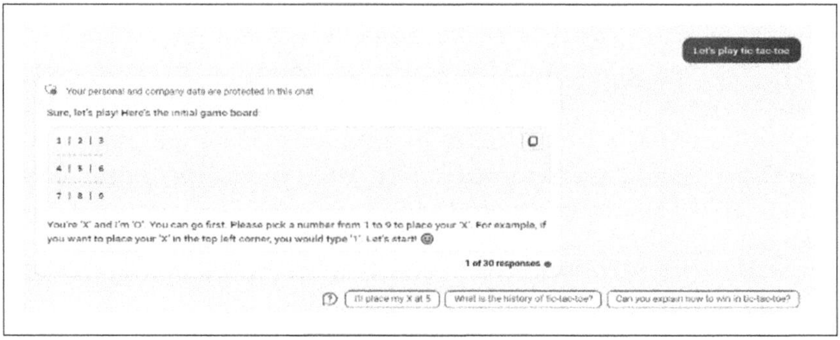

Figure 2.27 Tic-tac-toe (compiled by Copilot)

Creating images

Now let's move on to such an interesting feature as image creation using AI. To do this, you need to start a new conversation and ask a question, the answer to which should be an image.

Image Creator is different from searching for an image in Bing. It works best when you're highly descriptive. So, get creative and add details: adjectives, locations, even artistic styles such as "digital art" and "photorealistic." Here's an example: instead of a text prompt of "creature"—try submitting a prompt for "fuzzy creature wearing sunglasses, digital art" (Microsoft, n.d.).

Copilot (Figure 2.29) and Bing Image Creator (Figure 2.30) differ from each other in interface.

Bing Image Creator generate four image variants.

Create vector illustrations

In order to get the result that the user want, the user can explicitly ask Copilot to draw a specific image in the query field:

Prompt:
```
Draw a caricature of an office worker standing in front
of a dashboard with chart, blue and yellow colors,
use gradients.
```

COPILOT AND EDUCATION

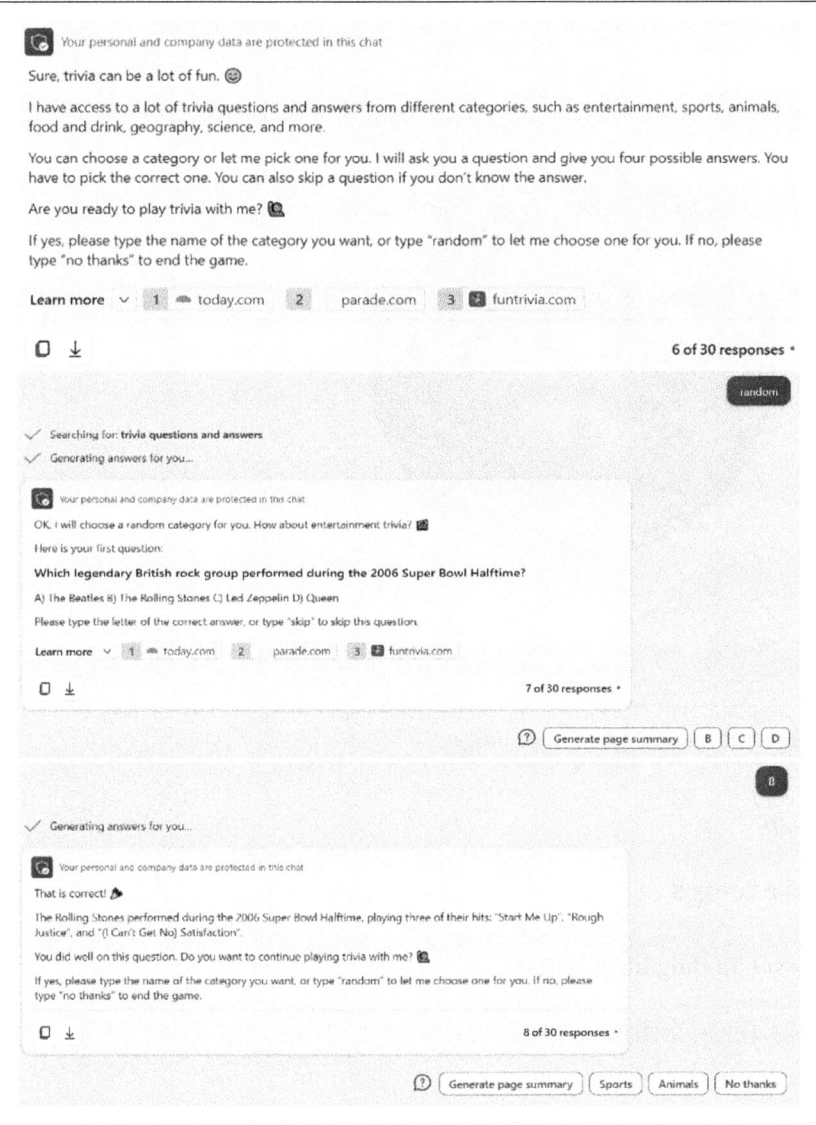

Figure 2.28 Trivia (compiled by Copilot)

CHAPTER 2

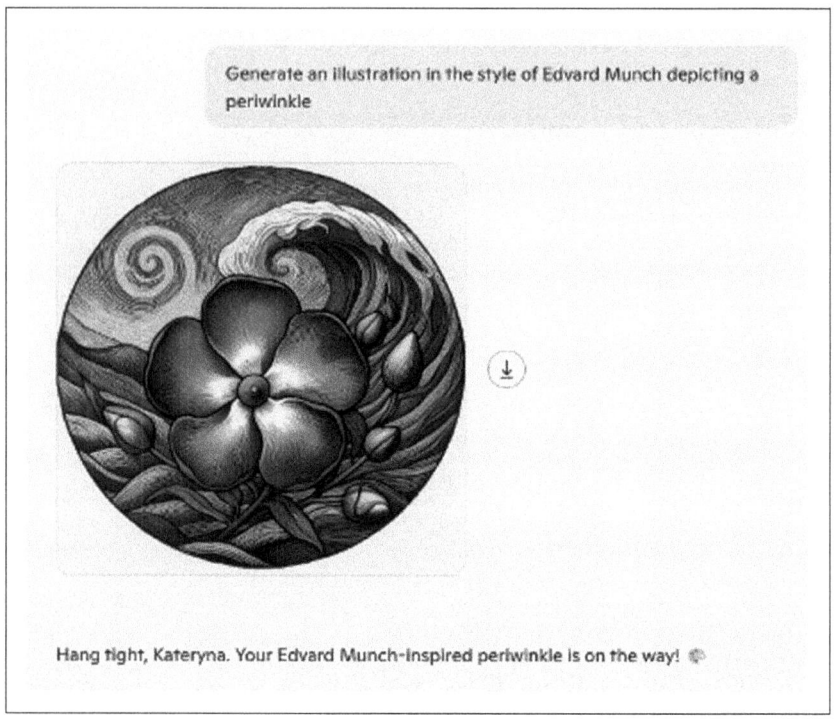

Figure 2.29 Copilot generates an image

Result:

Make comics

Prompt (in Bing Image Creator):
```
Three-panel comic strip that shows how Alan Turing
breaking ENIGMA
```

COPILOT AND EDUCATION

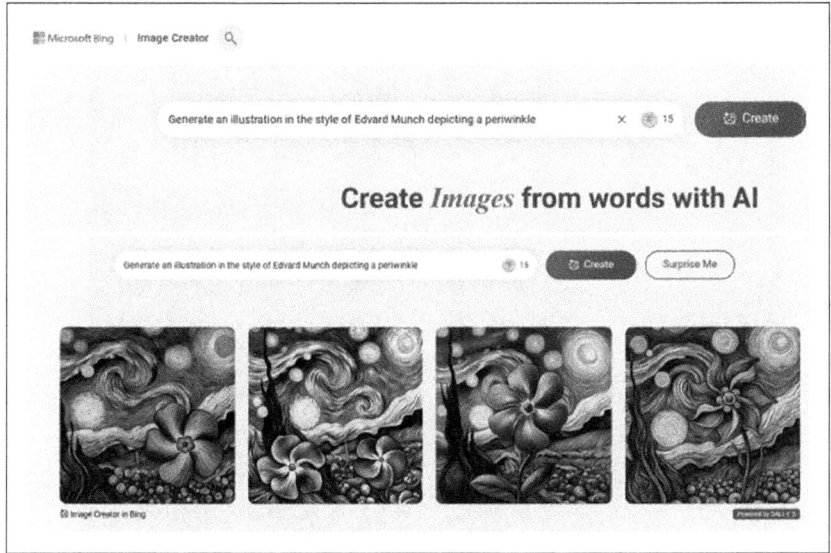

Figure 2.30 Bing Image Creator generates an image

Result:

Make logo

Prompt (in Bing Image Creator):
```
Create a logo using runes that represents a bookstore
```

Result:

Writing a dialogue

Prompt:
```
Write a dialogue between a buyer and a seller of a
television.
```

Figure 2.31 Bing Image Creator using "boost" or "boosted generation" (Figure 2.31).

Result:
Sure thing! Here's a dialogue between a buyer and a seller negotiating the purchase of a television:

Seller*:* Good afternoon! Welcome to our electronics store. How can I assist you today?

COPILOT AND EDUCATION

Figure 2.32 "Boost" in Bing Image Creator

Figure 2.33 Three-panel comic strip that shows Alan Turing breaking ENIGMA

> Sure thing, Kateryna! Here's a creative take on a three-panel comic strip showing Alan Turing breaking the ENIGMA code:
>
> **Panel 1: The Challenge**
>
> Scene: A dimly lit room filled with chalkboards covered in complex equations, and a large, intimidating machine (the ENIGMA machine) sits in the center. Text Box: "In the midst of World War II, a brilliant mathematician named Alan Turing was faced with an extraordinary challenge."
>
> **Panel 2: The Breakthrough**
>
> Scene: Alan Turing sits at a desk, surrounded by papers and codebooks. His eyes light up with a eureka moment, and he begins writing frantically. Text Box: "Turing's genius led to a groundbreaking discovery. He developed a machine, the Bombe, that could decode ENIGMA's complex cipher."
>
> **Panel 3: The Victory**
>
> Scene: The Bombe machine is whirring, and Turing stands triumphantly as the first decoded message is revealed. His colleagues cheer in the background. Text Box: "Turing's success in breaking the ENIGMA code played a crucial role in the Allied victory, saving countless lives and changing the course of history."
>
> I hope this helps! If you need more details or a specific visual element, just let me know.
>

Figure 2.34 Three-panel comic strip that shows Alan Turing breaking ENIGMA (a description)

Figure 2.35 Logo (compiled by Bing Image Creator)

> Here's a snippet of Python code that will add 5% to a user's number and then divide the result by 3:
>
> ```python
> def calculate_result(user_number):
>     # Add 5% to the user's number
>     increased_number = user_number * 1.05
>     # Divide the result by 3
>     final_result = increased_number / 3
>     return final_result
>
> # Example usage
> user_number = float(input("Enter your number: "))
> result = calculate_result(user_number)
> print("The result is:", result)
> ```
>
> This code defines a function `calculate_result` which takes a user's number as input, adds 5% to it, and then divides the result by 3. You can use the `input` function to get the user's number and print the result using the `print` function.
>
> If you try running it, let me know how it goes!

Figure 2.36 Python code

Buyer: Hi there! I'm looking to buy a new television. Do you have any recommendations?
Seller: Absolutely! We have a wide range of TVs. What size are you looking for, and do you have a preference for any specific brand or features?
Buyer: I'm thinking about a 55-inch TV. I'd like a 4K resolution and smart TV features. I'm not too particular about the brand, but I want something reliable and with good picture quality.
Seller: Great choice! Let me show you a few options. This model here is a 55-inch Samsung 4K UHD Smart TV. It has excellent picture quality, smart features like built-in streaming apps, and it's quite popular among our customers. It's priced at $799.
Buyer: That sounds good. Can you tell me more about the warranty and any ongoing promotions?

Seller: Sure! This TV comes with a one-year manufacturer's warranty. We also offer an extended two-year warranty for an additional $50. As for promotions, we have a 10% discount on all TVs purchased this week, and you'll also get a free HDMI cable.
Buyer: That's tempting. I'm also considering this LG model over here. How does it compare to the Samsung one?
Seller: The LG model is also a 55-inch 4K UHD Smart TV. It's slightly cheaper at $749, and it comes with similar smart features and picture quality. The main difference is in the operating system; the Samsung uses Tizen, while the LG uses webOS. Both are user-friendly, so it really comes down to personal preference.
Buyer: I see. The LG is a bit cheaper, but I've heard good things about Samsung's picture quality. Let me think about it for a moment.
Seller: Take your time! If you have any questions or need further assistance, I'm here to help.
Buyer: After some thought, I think I'll go with the Samsung. The picture quality and the overall features suit my needs better.
Seller: Excellent choice! I'll get that processed for you. Would you like to add the extended warranty?
Buyer: Yes, I'll take the extended warranty as well.
Seller: Perfect. I'll apply the discount and get you the free HDMI cable. Your total comes to $809. Please follow me to the checkout.
Buyer: Thank you! I'm excited to set this up at home.
Seller: You're welcome! Enjoy your new TV, and if you need any help with setup or have questions in the future, feel free to contact us.
Buyer: Will do. Have a great day!
Seller: You too!

```
I hope you found this dialogue helpful! If you need
anything else, feel free to ask.
```

Writing poems
Copilot can help the user write a poem, songs, or stories.

Prompt:
Write a poem about the advancements of artificial intelligence in the style of William Shakespeare.

Result:
    Of machinery and man's ambitious quest,
    When silicon and thoughts did intertwine,
    To raise aloft a being dispossessed,
    From gears and code a thinking spark align.
    What once was bound by flesh and bone, doth break,
    In circuits bright, the dawn of reason's light,
    A semblance of the human mind to make,
    And conquer thus the shadows of the night.
    With whispered code and algorithms' grace,
    A semblance of our wit it dost display,
    Yet lacking heart, nor feels it love's embrace,
    Cold logic guides it through the endless day.
    So marvel, humankind, at thine own art,
    Yet guard thy soul, lest thou forget its part.

2.3 AI and IT education: opportunities in pedagogical activity

One of the areas of IT education is the study of artificial intelligence technologies. Currently, generative artificial intelligence technologies are already used in IT education, and many scientists are raising questions about the goals, feasibility, usefulness, and ethics of using AI in IT education.

Concerning IT education, we can point out more narrow aspects concerning the specifics of professional training of IT specialists in universities. In particular, Becker et al. (2023) offers a number of avenues where AI-generated code tools present clear opportunities:

1) code solutions for learning—AI-generated solutions provide a low-cost way for students to generate exemplar solutions to check their work; code generation tools can be used to help expose students to the variety of ways that a problem can be solved; code generation models could be used to generate solutions of varying (or unknown) quality, and for assessment tasks focusing on the evaluation of code quality;
2) producing learning resources—AI could be used to programming exercises generation, code explanations and development of a large number of illustrative examples;
3) new pedagogical approaches—students using the AI-generated solutions can focus more on how to communicate algorithmic problems clearly; AI tools could help students get rid of the programmer's writer's block (fear of starting or continuing work) and are capable of explaining error messages in natural language.

Santos et al. assert that AI can support cognitive and motor development, stimulate reasoning by using such concrete cases of existing applications as games of intellect or strategy gaming (e.g., chess, Cluedo, Go) and gamification action games (e.g., Duck Hunt) (Santos et al., 2023). AI-based computer application the technology teaching can improve students' concentration and enthusiasm for learning (Li, 2022). Research has shown that deep learning as technology of AI can assist in automated assessments, helping students identify their flaws in learning computer programming (Shahidatul & Adidah, 2021). Additionally, the use of AI reduces workloads by shortening the time spent for assignments grading, preparing lesson plans, and other paperwork (Vinutha, Kavyashree, Vijay, & Raju, 2022).

Several tools based on AI technology can be used in education. One example is GitHub Copilot (<http://www.helloworld.cc/copilot>), which is an AI pair programmer that runs in an IDE (such as Visual Studio Code). Notably, "AI in programming isn't some far-off sci-fi dream anymore. With tools like Visual Copilot, you can convert Figma designs into code in one click or prototype features in a minute using Cursor" (Gopinath, 2024). Moreover, it is important to note that when the opportunity arises, Copilot predicts what's coming next and suggests it in code as ghost text. Such technology can be used to provide suggestions for improving a piece of code, generate a summary of a code change, or propose a strategy to solve an issue. This is useful in educational as well as professional settings (Rice, 2023). AI can be used to prevent cheating in programming through the implementation of online proctoring systems, for example, Proctorio and SMOWL. However, these systems still need improvement (Bergmans, Bouali, Luttikhuis, & Rensink, 2021). AI algorithms can furthermore be used to analyze students' behavior, such as in online proctoring systems for webcam detection of behavioral signs of fraud (Nigam, Pasricha, Singh, & Churi, 2021).

AI-powered plagiarism detection tools can prevent cheating, for example, by scanning students' written work and flagging potential cases of academic dishonesty. Tools that can be used for this purpose include Originality.ai, GPTZero, AI Detector Pro and Winston.ai. A tool called Giant Language Model Test Room (GLTR) (<https://gltr.io>) highlights passages that might have been generated by a computer program. Turnitin (<https://www.turnitin.com>) has developed several features to help teachers, in particular, the company launched AI detection to help educators identify when student submissions use AI writing tools like ChatGPT.

Thus, AI technologies are widely and variously used in IT education (Osadcha & Krogstie, 2023).

Copilot can also make code for different computer languages that many of us use, like Python, Java, C++, C#, and JavaScript.

Prompt:
Write Python code that will add 5% to the use's number, and then divide by 3.

Result:
In the same way, we can get code in Java, for example.

Bibliography

Abdullahi, A. (December 11, 2023). 20 best AI chatbots in 2024. Retrieved February 20, 2024, from <https://www.eweek.com/artificial-intelligence/best-ai-chatbots>.

AJE's Free Grammar Check for Scientists and Academics (2025). <https://www.aje.com/arc/free-grammar-check-for-scientists-and-academics/>

Almekinders, S. (2023, April 3). Google Bard PaLM update: 540 billion parameters take fight to ChatGPT. Retrieved October 19, 2023, from Techzine Europe: <https://www.techzine.eu/news/applications/104043/google-bard-palm-update-540-billion-parameters-take-fight-to-chatgpt>.

Becker, B. A., Denny, P., Finnie-Ansley, J., Luxton-Reilly, A., Prather, J., & Santos, E. A. (2023). Programming is hard - or at least it used to be: educational opportunities and challenges of AI code generation. Proceedings of the 54th ACM Technical Symposium on Computer Science Education V. 1. Association for Computing Machinery, New York, NY, USA, 500–506. <https://doi.org/10.1145/3545945.3569759>.

Bergmans, L., Bouali, N., Luttikhuis, M., & Rensink, A. (2021). On the efficacy of online proctoring using Proctorio. Proceedings of the 13th International Conference on Computer Supported Education, 1, 279–290. <https://doi.org/10.5220/0010399602790290>.

DuBay, W. H. (2004). The Principles of Readability. <https://files.eric.ed.gov/fulltext/ED490073.pdf>

Ebiner, P. (n.d.). ChatGPT, Midjourney, Firefly, Bard, DALL-E, AI crash course. Retrieved October 19, 2023, from Udemy: <https://www.udemy.com/course/chatgpt-midjourney-google-bard-dall-e-ai-course>.

Edwards, B. (2023, May 11). The AI race heats up: Google announces PaLM 2, its answer to GPT-4. Retrieved from ArsTechnica: <https://arstechnica.com/information-technology/2023/05/googles-top-ai-model-palm-2-hopes-to-upstage-gpt-4-in-generative-mastery>.

GenAI Chatbot Prompt Library for Educators. (n.d.). Retrieved November 14, 2023, from AI for Education: <https://www.aiforeducation.io/prompt-library-lesson-planning>.

Goode, L. (2023, March 30). Review: we put ChatGPT, Bing Chat, and Bard to the test. Retrieved October 19, 2023, from WIRED: <https://www.wired.com/story/review-ai-chatbots-bing-bard-chat-gpt>.

Gopinath, V. (2024). AI Pair Programming in 2025: The Good, Bad, and Ugly. <https://www.builder.io/blog/ai-pair-programming>

Hussain, S. (2022, December 22). Get to know Midjourney AI art: an introduction. Retrieved from Medium: <https://medium.com/cryptocurrencies-ups-and-down/get-to-know-midjourney-ai-art-an-introduction-2698344078bb>.

Introducing the New Bing. (2023, October 6). Retrieved from Bing: <https://www.bing.com/new?setlang=en&sid=02D9A5B81B6B66992127B6331A5B6737>.

Kaelin, M. W. (2023, June 7). 8 tips for achieving better results from the new Bing AI. Retrieved December 12, 2023, from Tech Republic: <https://www.techrepublic.com/article/bing-chat-ai-tips-for-better-results>.

Kourkoulou, D. (2023). Replika AI: technological affect and general AI imaginations. The International Journal of Communication and Linguistic Studies, 21(2), 73–86. <https://doi.org/10.18848/2327-7882/CGP/v21i02/73-86>.

Li, P. (2022). Teaching transformation of computer application technology specialty based on artificial intelligence. In Hung, J. C., Chang, JW., Pei, Y., & Wu, W. C. (Eds.), Innovative computing. Lecture Notes in Electrical Engineering. Vol. 791. Springer.

Licuan, C. (2023, September 21). How to use Bing AI chat in 2023 (Beginner's Guide). Retrieved October 18, 2023, from Elegant: <https://www.elegantthemes.com/blog/business/how-to-use-bing-ai-chat>.

Lisowski, E. (2023, March 23). Bing Chat vs. ChatGPT. Which is better? Retrieved March 15, 2025, from: <https://addepto.com/blog/is-bing-chat-the-same-as-chatgpt-main-differences/>

Loukides, M. (2023). What are ChatGPT and its friends? O'Reilly Media, Inc.

Mariciuc, D. F. (2023). A bibliometric analysis of publications on customer service chatbots. Management Dynamics in the Knowledge Economy, 11(1), 48–62.

Mauran, C. (2023). Meet Copilot, Microsoft's AI tool for work and productivity. <https://mashable.com/article/microsoft-copilot-ai-assistant-announcement>.

Mehdi, Y. (2023, September 21). Announcing Microsoft Copilot, your everyday AI companion. Official Microsoft Blog. <https://blogs.microsoft.com/blog/2023/09/21/announcing-microsoft-Copilot-your-everyday-ai-companion>.

Microsoft. (n.d.). Image Creator from Designer. Retrieved November 14, 2023, from: <https://www.microsoft.com/en-us/edge/features/image-creator?form=MT00D8>.

Microsoft. (2023). What will you do with Copilot in Bing? From Microsoft: <https://www.microsoft.com/en-us/bing?ep=251&es=31&form=MA13FV>.

Microsoft. (2023, February 7). Reinventing search with a new AI-powered Microsoft Bing and Edge, your copilot for the web. Official Microsoft Blog. Retrieved November 6, 2023, from <https://blogs.microsoft.com/blog/2023/02/07/reinventing-search-with-a-new-ai-powered-microsoft-bing-and-edge-your-copilot-for- the-web>.

MS Copilot Chat - Guide for IT Educators. (2024). Retrieved from <NTNU.EDU>: <https://www.ntnu.edu/excited/ms-copilot-chat-guide-for-it-educators>.

Nigam, A., Pasricha, R., Singh, T., & Churi, P. (2021). A systematic review on AI-based proctoring systems: past, present and future. Education and information technologies, 26(5), 6421–6445. <https://doi.org/10.1007/s10639-021-10597-x>.

Official Microsoft Blog, The (2023). Introducing Microsoft 365 Copilot—your copilot for work. <https://blogs.microsoft.com/blog/2023/03/16/introducing-microsoft-365-copilot-your-copilot-for-work/>.

OpenAI. (n.d.). Models—OpenAI API. Retrieved October 20, 2023, from Platform OpenAI: <https://platform.openai.com/docs/models/gpt-4>.

OpenAI. (2023, March 14). Microsoft Copilot [Large language model]. <https://sl.bing.net/ezlCgyftXy0>.

OpenAI. (2023, August 22). GPT-3.5 Turbo fine-tuning and API updates. Retrieved October 22, 2023, from OpenAI: <https://openai.com/blog/gpt-3-5-turbo-fine-tuning-and-api-updates>.

Ortiz, S. (2023, June 2). 7 ways you didn't know you can use Copilot and other AI chatbots. Retrieved December 14, 2023, from ZDNET: <https://www.zdnet.com/article/7-ways-you-didn>'t-know-you-can-use-bing-chat-and-other-ai-chatbots.

Osadcha, K. & Krogstie, B. R. (2023). Ways of using artificial intelligence in IT education of Norway NIKT: Norsk IKT-konferanse for forskning og utdanning. No. 4. <https://www.ntnu.no/ojs/index.php/nikt/article/view/5711>.

Osadchiy, V., & Osadcha, K. (2024). Використання Microsoft Copilot у вищій освіті та наукових дослідженнях: навчально-методичний посібник. Київ: ІЦО НАПН України. 90 с.

Plevris, V., Papazafeiropoulos, G., & Jiménez, R. A. (2023). Chatbots put to the test in math and logic problems: a comparison and assessment of ChatGPT-3.5, ChatGPT-4, and Google Bard. AI, 4(4), 949–969. <https://doi.org/10.3390/ai4040048>.

Purtill, J. (2023, April 1). Prompt engineers share tips on using ChatGPT, Midjourney, Copilot and other generative AI. Retrieved October 19, 2023, from ABC: <https://www.abc.net.au/news/science/2023-04-02/prompt-engineers-share-their-tips-on-using-chatgpt-generative-ai/102165132>.

ProWritingAid. (2025). <https://prowritingaid.com/readability-checker>

Rebelo, M. (2023, August 14). The best AI chatbots in 2023. Zapier. Retrieved November 8, 2023, from <https://zapier.com/blog/best-ai-chatbot>.

Rice, A. (2023, April 7). Using AI-powered developer tools for teaching programming—Hello World. Hello World. Retrieved August 11, 2023, from <https://helloworld.raspberrypi.org/articles/hw20-using-ai-powered-developer-tools-for-teaching-programming>.

Role of Grammar in Enhancing Readability and Clarity (2025). <https://www.wordsdetail.com/role-of-grammar-in-enhancing-readability-and-clarity/#google_vignette>

Rosset, C. (2020, February 13). Turing-NLG: A 17-billion-parameter language model by Microsoft. Retrieved October 18, 2023, from Microsoft Research Blog: <https://www.microsoft.com/en-us/research/blog/turing-nlg-a-17-billion-parameter-language-model-by-microsoft>.

Rudolph, J., Tan, S., & Tan, S. (2023). War of the chatbots: Bard, Bing Chat, ChatGPT, Ernie and beyond. The new AI gold rush and its impact on higher education. Journal of Applied Learning & Teaching, 364–389.

Santos, V., Mamede, H., Silveira, C., & Reis, L. (2023). A reference model for artificial intelligence techniques in stimulating reasoning, and cognitive and motor development. Procedia Computer Science, 219, 1057–1066. <https://doi.org/10.1016/j.procs.2023.01.384>.

Sha, A. (2023, March 20). 15 best ways to use Bing AI. Retrieved December 14, 2023, from Beebom: <https://beebom.com/best-ways-use-bing-ai>.

Shahidatul, A. B., & Adidah, L. (2021). Deep learning approach for cognitive competency assessment in computer programming subject. International Journal of Electrical and Computer Engineering Systems, 12, 51–57. <https://doi.org/10.32985/ijeces.12.si.6>.

Siegle, D. (2023). A role for ChatGPT and AI in gifted education. Gifted Child Today, 46(3), 211–219. <https://doi.org/10.1177/10762175231168443>.

Singh, N. (2023, August 3). ChatGPT vs Copilot vs Bard. <https://verloop.io/blog/generative-ai-comparison>.

Stefanowicz, B. (2024, January 29). 22 Best AI chatbots for 2024: ChatGPT & alternatives. Retrieved February 20, 2024, from <https://www.tidio.com/blog/ai-chatbot>.

Studio, N. (2022, May 23). How does NightCafe AI work? NightCafe Creator. Retrieved October 19, 2023, from NightCafe Creator: <https://nightcafe.studio/blogs/info/how-does-nightcafe-ai-work>.

Ventoniemi, J. (n.d.). What is an AI chatbot? Here's what you need to know (+Infographics). Giosg. Retrieved November 6, 2023, from <https://www.giosg.com/blog/what-is-ai-chatbot>.

Vinutha, D. C., Kavyashree, S., Vijay, C. P., & Raju, G. T. (2022). Innovative practices in education systems using artificial intelligence for advanced society. In S. K. Panda, R. K. Mohapatra, S. Panda, & S. Balamurugan (Eds.), The New Advanced Society: Artificial Intelligence and Industrial Internet of Things Paradigm (pp. 351–372). John Wiley & Sons, Incorporated. <https://doi.org/10.1002/9781119884392.ch16>.

Vynnychuk O. T. (2020). Pedagogical management is a theory and practice of managing the educational process. Actual problems of management of academic institutions in the context of the strategy of modernization of the educational sector. Ternopil: TNPU is named after V. Gnatiuk, P. 19–30.

… CHAPTER 3

AI and Foreign Language Education: The Theoretical and Practical Dimensions

Language education today can be carried out using AI-powered technologies. Generative AI tools provide potential for language acquisition. Since late 2022, AI-powered technology, such as ChatGPT and similar large language models (LLMs), has appeared in the educational space.

AI technology is a resource that provides unprecedented, sometimes phenomenal, opportunities to enhance both efficiency (improve pronunciation, contribute to developing writing skills) and accessibility (free online resources; paid resources that are universally accessible online) to foreign language learning.

The integration of AI in language education rests on the concept of constructivism. AI technology enables simulating real-life conversations, offering instant feedback, and adapting to individual learners' pace and language proficiency.

The practical application of AI technology (intelligent tutoring systems, conversational agents, and gamified language exercises) is gradually making its contribution to foreign language acquisition by learners. AI tools, by tutors' supervision, of course, help overcome language barriers (poor communication skills, accents), resulting in more effective learning outcomes.

However, the adoption of AI in language education is not without difficulties. Researchers, teachers, and learners actively discuss ethical issues related to AI.

This chapter explores the theoretical props and practical underpinnings required to learn and teach a foreign language with AI technology involvement.

3.1. What do scientists say? Faced with compelling new theories on foreign language learning and teaching

Recent publications related to the use of AI in language education shed light on many theoretical and practical aspects of teaching and learning foreign languages (Table 3.1).

Currently, there are many AI tools for teaching and learning foreign languages. The use of AI tools for knowledge assessment (pronunciation analysis, text analysis (plagiarism, cohesion), grammar checkers, vocabulary analysis) and teaching (chatbots, language-learning platforms) foreign languages contribute or aim to improve the language proficiency of learners. In the recent research of Son et al. (2023), AI technologies and applications that exist today are described. The researchers distinguished AI tools and applications into categories that were designated based on AI technologies' functionalities, applications, and modes of interaction within the language classroom. Considering the research findings of Son et al. (2023), we expended the list of categories and then provided a characteristic of each one. Below is the list of the categories: (1) ASR and pronunciation AI tools; (2) AI-driven language-learning platforms; (3) virtual language assistants and chatbots; (4) NLP; (5) AI-enabled adaptive learning systems; (6) CDA; (7) AI-driven language translation tools; and (8) AWCF offered by Grammarly. A characteristic of each of the categories mentioned above is given below.

Category one: automatic speech recognition (ASR) and pronunciation AI tools. ASR is AI technology that employs AI power and machine learning techniques to determine, comprehend, and decode spoken speech and written text. Here are some examples. ELSA Speak (<elsaspeak.com>)—an AI-powered English-speaking coach that helps recognize the speech patterns of non-native speakers. It enables real-time speech recognition feedback on pronunciation and fluency. Another example is Speechify (<speechify.com>)—a text-to-speech reader and AI voice generator. It is used in voice recognition and speech-to-text software applications, such as intelligent personal assistants (IPAs), automatic transcribers, and notetaking apps (Evers & Chen, 2020). ASR is also used on smartphones to dictate a message to a phone. The phone understands the language and performs an action using the language. ASR has progressed rapidly over the last decade, becoming more accurate

Table 3.1 Key findings, best practices, key points from publications on the research on AI in language education

No.	The author(s)	Key findings, best practices, key points
1	Jeon (2022)	The article is devoted to exploring AI chatbot affordances in the EFL classroom. Based on the abstract information available online, the findings confirm that "depending on the learner, the chatbot affordances were perceived differently; thus, each affordance acted as either an opportunity or a constraint for English language learning" (Jeon, 2022).
2	Liu and Hung (2016)	The use of AI contributes to improving learners' pronunciation. The authors stated: "The results of the analysis on learner scores rated by MyET show that the participants' pronunciation performance of the prompted sentences was significantly improved after the training" (Liu and Hung, 2016).
3	Dizon and Gold (2023)	The article is devoted to studying the effects of AWE on L2 students. The paper explored the impact of Grammarly, a popular AWE tool, on EFL students' FLA and LA. The study's results showcased that Grammarly significantly positively affected both FLA and LA.
4	Fu et al. (2022)	The article is devoted to reviewing AWE feedback (types, learning outcomes, implications). Authors stated: "AWE feedback to some extent can improve students' writing from the product-oriented aspect but is not as effective as human feedback (e.g., teacher or peer feedback)" (Fu et al., 2022).
5	Kazu and Kuvvetli (2023)	The research findings confirmed that pronunciation practice via AI-enabled the words to remain in memory longer. Researchers wrote: "Based on the interview carried out at the end of the process, it was mentioned that the majority of the participants had benefited from pronunciation training with artificial intelligence-supported speech recognition and that it contributed to them in various areas" (Kuzu & Kuvveli, 2023).

No.	The author(s)	Key findings, best practices, key points
6	Lee et al. (2023)	Scholars explored an LGC approach (this approach is characterized as the creation and exploit of digital technology that allows learners to build a "learner-generated context" and acquire knowledge within it). According to the researchers, the LGC—an AI-powered pedagogical approach—encourages and fosters learners' self-autonomous learning activities. Researchers drew on the LGC framework and developed research methodology, developed an AI-based English learning support system for Korean learners and examined whether and how the system could catalyze forming LGC-based learning experiences among the learners they studied. From their findings, they argued that "AI-applied learning assistance system based on sound educational technological design framework (specifically, in this study, the LGC framework) can catalyze learners' autonomous learning experiences, even without a specified instructor, curriculum, or location, and help them become creators of their learning contexts. This study also provides a reference for AIEd researchers and practitioners pursuing similar goals" (Lee et al., 2023).
7	Fryer et al. (2020)	Researchers explored the potential of using (chat) bots for language learning. In the authors' view, "chatbots are a new, revolutionary stage for foreign language learning" (Fryer et al., 2020). Chatbots' potential role in supporting critical components of interaction competence is studying. The potential usefulness of Cleverbot (quote from the article: "Cleverbot has its own brand of Artificial Intelligence software, a key concept of which is context," Fryer et al., 2020) and Mondly (quote from the article: "Mondly chatbots were developed as part of a language-learning platform, rather than as a means of entertaining human users as was Jabberwacky/Cleverbot," Fryer et al., 2020).

No.	The author(s)	Key findings, best practices, key points
8	Illingworth (2023)	The publication is devoted to the thoughts on ChatGPT. The author reflects on the implementation of ChatGPT in the field of tertiary education. He considered the challenges related to the use of ChatGPT. ChatGPT's capabilities offer opportunities to improve assessment at the higher education level by promoting more authentic and relevant tasks that mirror real-life challenges. However, it also brings challenges and concerns (potential bias in grading, ethical dilemmas about cheating, and the risk of perpetuating systemic inequalities if misused).
9	Liu, Bridgeman, and Miller (2023)	ChatGPT is the point of discussion in this publication. It can save teachers' time and help students learn. The opportunities provided by ChatGPT are considered. ChatGPT and similar AI tools present opportunities to enhance teaching and learning by streamlining lesson preparation, generating discussion prompts, and providing personalized support for students. Also, traditional assessments face challenges due to their ability to produce passable written outputs. The authors believe that teachers should use AI to create educational materials, engage students in critical discussions about AI's strengths and biases, and adapt assessments to focus on higher-order thinking and contextualization. In the researcher's view, students should use AI responsibly to improve their writing content or gain feedback while focusing on their learning process. Prioritizing teacher feedback can help track and control student progress and reduce dependence on AI tools. Collaboration between educators and students is essential to integrate AI into teaching–learning effectively.

No.	The author(s)	Key findings, best practices, key points
10	Loble (2023)	ChatGPT as an example of technology is studied by the author. The case of Australia is under research. The author wrote: "The range of responses to ChatGPT shows how education has yet to figure out the best way to use such tool" (Loble, 2023). Summary: Governments should establish quality standards for edtech, provide professional development for teachers, and regulate data collection and usage to protect student and teacher privacy. Schools should prioritize tools grounded in proven teaching methods and independently evaluated for effectiveness. Key findings: The global edtech market is expanding, with major competitors like Google and Microsoft and emerging investments from China and the EU. Effective implementation depends on how well schools integrate these tools into curricula and adapt them to teachers' work.
11	Pérez-Paredes (2022)	The study reviewed research on data-driven learning (DDL) and corpus use in language teaching from 2011–2015 in five CALL-related journals, finding limited DDL normalization primarily in contexts where researchers and teachers shared roles, particularly in Asia, Europe, and the US. The author identified two key areas that require attention: (1) syllabus integration and (2) language teacher training.
12	Koltovskaia (2023)	Research findings claim that when used to complement teacher feedback, Grammarly did not create a division of labor between addressing higher-order (global) and lower-order (local) writing concerns, as teachers provided feedback on both aspects. Effective use of Grammarly as a complement requires thoughtful integration into teacher's work and feedback provided by the teacher, using, relying on and often trusting Grammarly's strengths (i.e., correctness, clarity, engagement).

No.	The author(s)	Key findings, best practices, key points
13	Barrot (2021)	The study, as can be seen from the abstract (available online), highlights the effectiveness of automated written corrective feedback (AWCF) via Grammarly in improving college students' writing accuracy, particularly by promoting noticing and encouraging self-directed learning through adaptive metalinguistic explanations. The author defined challenges (overcorrection, cognitive overload, and limited metalinguistic explanations), indicating areas for improvement in AWCF systems. AWCF's ability to adapt to error severity levels and engage students in reflective learning suggests its potential as a valuable tool in writing pedagogy. According to the researcher's view, to maximize effectiveness, tutors should balance AWCF use with strategies to mitigate its limitations, ensuring learners are not overwhelmed.
14	Wang et al. (2022)	The study, as can be seen from the abstract (available online), reveals that the involvement of AI in the L2 learning process, more precisely the application of an artificial intelligence (AI) coach for second language (L2) learning, contributes to enhancing learners' outcomes. Study findings highlight the value of developing AI tools with engaging features to optimize the effectiveness of educational content.
15	Khurana et al. (2023)	The researchers distinguished four phases by discussing different levels of natural language processing (NLP) and components (NLP can be classified into two parts: Natural Language Understanding and Natural Language Generation) of NLP. The authors schematically presented the broad classification of NLP (natural language processing → **natural language understanding** (it involves: phonology, morphology, pragmatics, syntax, semantics) + **natural language generation** (it involves: natural language test):

No.	The author(s)	Key findings, best practices, key points
		Natural Language Processing ├──────────────┬──────────────┤ Natural Language Understanding Natural Language Generation
		* This schematic presentation of a broad classification of NLP is also showcased in Abro et al. (2023). Applications of NLP (machine translation, text categorization, spam filtering, information extraction, summarization, dialogue system, and medicine) were considered in detail. The researchers discussed the recent NLP projects (e.g., the future of BI in natural language processing) implemented by companies. Scholars gave a walkthrough of recent developments (by years) in NLP: neutral language model (2001), multitask learning (2008), word embedding (2013), neutral network for NLP (2013), sequence to sequence model (2014), attention mechanism (2015), and pretrained language (2018).
16	Chen et al. (2024)	The paper offers a comprehensive review of the applications and methodologies of AI in NLP. The paper also explores specific AI techniques used in NLP, including machine learning. Researchers descriptively characterized machine translation, chatbots, and virtual assistants.
17	Randall and Urbanski (2023)	The paper, as can be seen from the abstract (available online), discloses the functional characteristics of a computerized dynamic assessment (CDA) program that was created to help teachers develop learners' understanding of second language grammatical features. The program CDAG, as researchers called it, approximates an interactionist approach to dynamic assessment, using prompts calibrated to the learner's answer choice. The authors wrote, "CDAG allows the instructor to customize language, question type, linguistic features to be assessed, answer choices (including distractors), and mediational prompts: it can thus assess a variety of grammatical structures."

No.	The author(s)	Key findings, best practices, key points
18	Kargar Behbahani and Karimpour (2024)	The paper, as can be seen from the abstract (available online), is devoted to studying the transformative potential of computerized dynamic assessment (CDA) on students' implicit and explicit language knowledge of grammar, concentrating on past perfect tense. In the authors' view, "Grounded in sociocultural theory (SCT), CDA integrates assessment and mediation within the Zone of Proximal Development (ZPD), offering personalized support to learners. The interface position in language acquisition theory posits that explicit knowledge can evolve into implicit through extensive practice."
19	Mohamed et al. (2024)	Researchers conducted a comprehensive review to provide a nuanced understanding of the present-day state of AI-driven language translation. AI-based translation approaches are systemized and as follows: (1) statistical machine translation (SMT), (2) rule-based machine translation (RBMT), (3) example-based machine translation (EBMT), (4) neutral machine translation (NMT), (5) hybrid systems, (6) zero-shot translation, (7) multimodal translation, (8) meta-learning for translation, (9) domain-specific models, and (10) reinforcement learning for MT. The authors examined the concepts related to AI-driven language translation. The examined concepts are Machine Learning, Deep Learning, Statistical Machine Translation, Natural Language Processing, Neural Machine Translation, Fuzzy Algorithms, Feature Extraction, and Evaluation Metrics.
20	Kushmar et al. (2022)	The paper focuses on examining the role of AI English language learning. Scientists wonder if AI is effective for English language learning and what practical methods can be used to apply AI technology effectively. In researchers' view, "AI has the potential to transform the functioning of the education system, increase the competitiveness of institutions and empower teachers and students at all levels. With intelligent content of instructions and testing, artificial intelligence allows focusing on the needs of the students."

No.	The author(s)	Key findings, best practices, key points
21	Jinming and Ben Kei (2024)	The findings of the paper claimed that "AI chatbot learning approach for English speaking proficiency was intended to speed up the English learning process and assist students in meeting the goals or results of the courses."
22	Crompton et al. (2024)	The study examined specific challenges and affordances for exploiting AI in ELT. Also, the paper "provides the scholarly community with a unique systematic review in the use of AI in ELT/L across learner levels." Moreover, the benefits that AI provides for ELT/L and that are valuable for self-regulation, pedagogy, reading, writing, and speaking are identified. Scientists are not spared challenges related to this matter (technology breakdowns, limited capabilities, fear, standardizing language).
23	Färber and Popovic (2023)	The authors propose Vocab-Expander (<vocab-expander.com>), an online tool that enables the creation and expansion of a vocabulary in the users' domain of interest.
24	Peng et al. (2023)	Peng et al. (2023) designed and developed Storyfier via a two-phase process:

No.	The author(s)	Key findings, best practices, key points
		Figure: Two-phases design and development process of Storyfier with teachers, learners, and HCI researchers. The researchers clarified the mechanisms of employing Storyfier, an AI-supported learning tool. They explained that Storyfier is a vocabulary learning system that "facilitates users to master the meaning and usage of any target English words via AI-generated stories and writing assistance." Authors suggest pedagogical strategies and activities for vocabulary learning.
25	Shumeiko (2024)	The research focuses on exploiting the potential of AI in teaching English. The characteristics of AI-powered language learning apps are given. An overview of the benefits and challenges of using AI for learning English is provided in a graphical format.
26	Edmett at al. (2023)	Authors concentrate on the benefits and risks of using AI in learning English. The British Council's survey of teachers, conducted to gather their views on using AI in teaching and learning English, is analyzed.
27	Shumeiko and Spišiaková (2025)	The book chapter outlines two sets of information regarding the use of AI resources in teaching and acquiring EFL. "The first set presents a scientific perspective on the appropriateness of using ChatGPT, Diffit, Wordwall, and Twee in EFL instruction, with examples of practical applications provided. The second set highlights the capabilities of AI tools that support learners in the process of EFL acquisition. Tools such as ChatGPT with voice integration, Mizou, GetPronounce, Quizlet, and Duolingo function as effective AI-powered facilitators for language comprehension."

and widely implemented in various industries (Daniels & Iwago, 2017; Son et al., 2023). In a review of technology types and their effectiveness, Golonka et al. (2012) stated that the measurable impact of technology on FL learning largely came from studies on ASR.

Category two: AI-driven language-learning platforms. AI-driven language-learning platforms can help learners to build a foundation of vocabulary and grammar to help them learn to speak and read in foreign language. AI has contributed to how teachers teach and learners learn, particularly in language learning. By providing a personalized path that adapts to each learner's learning pace and needs, AI-powered platforms (e.g., Duolingo, Babble, Rosetta Stone, and Memrise) are transforming language education from a one-size-fits-all model to a unique learning journey.

Category three: virtual language assistants and chatbots. AI-driven virtual assistants engage trainees in conversational interactions, providing real-time feedback, language training, and assistance (Son et al., 2023). Let us give examples: ChatGPT and Kuki (chat.kuki.ai). The conducted analyses of the publications of Coniam (2014) and Wang et al. (2021) support the notion that a chatbot is a software application that interacts with users via chat. A chatbot is also known as a bot, a virtual or a conversational agent, a chatterbot, a dialogue system, a virtual assistant, and, based on the thoughts of Kim et al. (2021), stimulates human conversations by asking and answering questions via text or audio. Relying on the characteristics provided by Fryer et al. (2020) and Wang et al. (2021), chatbots are commonly found on companies' websites in a range of industries, such as marketing, healthcare, technical support, customer service, and education, providing targeted services to website visitors. As Kim et al. (2021) and Smutny and Schreiberova (2020) explained, commonly, a user asks the chatbot a question, then the chatbot interprets the input, processes the user's intent, and only then provides a programmed response to the user. Chatbots commonly perform form-filling tasks such as collecting information to confirm someone's identity or information about a problem or an item they want to purchase and then directing them to an answer or preparing the information for a human to review quickly and easily.

ChatGPT (<https://chat.openai.com/>) has recently generated significant interest in various fields. Taking into consideration the opinions of Vincent (2022), Illingworth (2023), Liu et al. (2023), and Loble (2023), ChatGPT produces detailed written responses to requests for information based on vast databases. While ChatGPT has a substantial issue with factual accuracy, many tutors and researchers discuss its impact on education. In Zhai's (2022) view, ChatGPT can be used in writing an academic paper. The author believes that the text written by the AI chatbot was readable and informative and suggested that improving learners' creativity and critical thinking should be concentrated in education. ChatGPT might offer tempting opportunities for language teachers to enhance language teaching and create an engaging language training process for their trainees if carefully planned and used. Researchers hold the opinion that "GPT-4 and GPT-3.5 are the two most widely used large language models (LLMs)" (Chen et al., 2024).

Category four: natural language processing (NLP). According to Khurana et al. (2023), NLP has gained attention for exploring and studying human language computationally. NLP is classified into Natural Language Understanding (NLU) and Natural Language Generation (NLG). Khurana et al. (2023) wrote: "Linguistics is the science which involves the meaning of language, language context and various forms of the language." The all-encompassing classification of NLP includes the following notions: NLU (phonology, morphology, pragmatics, syntax, semantics) and NLG (natural language text).

Chen et al. (2024) have devoted their scientific paper to exploring AI methods in NLP. These researchers, highlighting the importance of AI in NLP, stated that "AI is a multidisciplinary field that aims to create intelligent machines capable of mimicking human cognitive functions." They define NLP in this way: "NLP is a specialized area within AI that focuses on enabling machines to understand, interpret, and generate human language." It is worth noting that NLP tasks, such as text summarization and document translation, rely on AI technologies. It is also important to emphasize that machine learning and deep learning are essential for handling the complexities of human language (with idioms and context-dependent information).

Category five: AI-enabled adaptive learning systems. Through detailed systematic mapping of the literature on AI-enabled adaptive learning systems, Kabudi et al. (2021) concluded that "systems (adaptive learning systems, intelligent mechanisms and adaptive learning platform) and frameworks for adaptive learning were the most proposed and utilised interventions for addressing the challenges faced by students and teachers." As this team of scientists stated, the importance of AI-enabled adaptive learning systems has increased during the pandemic as these systems can assist pedagogues in maintaining high-quality teaching and learning.

Category six: computerized dynamic assessment (CDA). In their paper, Kargar Behbahani and Karimpour (2024) highlight the potential of CDA "to individualize instruction, informing evidence-based language education policies." The study of Randall and Urbanski (2023) concentrates readers' attention on the CDA program, namely CDAG (as the authors called it), which was "created to help instructors develop student understanding of second language grammatical features and as a diagnostic and assessment tool. The program (called CDAG) approximates an interactionist approach to dynamic assessment, utilizing hints/prompts calibrated to the student's answer choice. CDAG allows the instructor to customize language, question type, linguistic features to be assessed, answer choices (including distractors), and mediational prompts: it can thus assess a variety of grammatical structures." In a paper by Ebadi and Saeedian (2016), the effects of computerized dynamic assessment (CDA) on promoting at-risk advanced Iranian EFL learners' reading skills are studied. The data was collected using the instruments of the DIALANG software and the Computerized Dynamic Reading Test (CDRT), developed by Prishghadam and Barabadi (2012). DIALANG (<http://dialangweb.lancaster.ac.uk>) is a free online assessment system that is proposed for individual language learners who want to obtain diagnostic info about their proficiency within the following skills: Reading, Listening, and Writing, and two more subskills (Grammar and Vocabulary) for fourteen languages. As Ebadi and Saeedian (2016) mentioned, "DIALANG's Assessment Framework and self-assessment statements are based on the CEFR; thus, it also gives feedback on the strengths and weaknesses of the learner's proficiency and advises about how to improve language skills."

In a study by Qin and Zhang (2019), dynamic assessment (DA) is presented as a resource for diagnosing language-related issues. Also, DA, to the authors' view based on previously published scientific papers, has been implemented to promote L2 learners' development. It is worth noting that DA, as mentioned in Qin and Zhang's (2019) paper, is used in educational and psychological assessment to assess the cognitive modifiability of the test takers.

In Kamrood et al.'s (2021) view, "Dynamic Assessment (DA) is proposed as a workable diagnostic tool in a second or foreign language context. Compared to traditional non-dynamic testing, DA presents a more comprehensive account of human beings' abilities through addressing both the fully internalized abilities and the abilities that are in the process of being internalized."

Category seven: AI-driven language translation tools. AI-driven language translation tools are software programs that automatically exploit AI techniques to translate speech or text from one language to another. There are reasons why people use AI-driven language translation tools. Here are some of them. AI-driven language translation tools can translate text faster than people can. They are available online. So, anyone with Internet access can translate the text. Free AI-driven language translation tools are Google Translate, DeepL, ChatGPT, Bing Microsoft Translator, Smartling, MachineTranslation.com, Alexa Translation, Taia, Mirai Translate, Reverso, Sonix, and Systran Translate Professional.

Category eight: automated written corrective feedback (AWCF) offered by Grammarly.

In a study exploring postsecondary teachers' use and perceptions of Grammarly (<https://www.grammarly.com/>), Koltovskaia (2023) noted that automated writing evaluation (AWE) can liberate educator's time to concentrate efforts on another or more serious issues. AWE is a software program that provides instant automated scoring and feedback for text improvement. It is worth noting that Grammarly, being an AWE, can be used as a complement to pedagogue feedback, but not replace pedagogue's feedback.

According to Barrot (2021), "the differential effects of automated written correction feedback (AWCF) on errors with different severity levels and gains across writing tasks remain unclear," compared to more or less previously studied automated writing evaluation (AWE) systems. The author explored how AWCF (through Grammarly) affects learners' writing accuracy. The results of this study approved the value of AWCF in providing adaptive metalinguistic description or explanation, along with engaging learners in self-directed learning. The study also discloses some challenges (i.e., cognitive overload and limited metalinguistic explanation). Barrot's research accepted the following AWCF model (that was primarily drawn on the interaction theory of Long, 1996):

Dizon and Gold (2023) investigated if Grammarly (an AWE tool) significantly impacted Japanese EFL students' FLA and LA. Results confirmed that

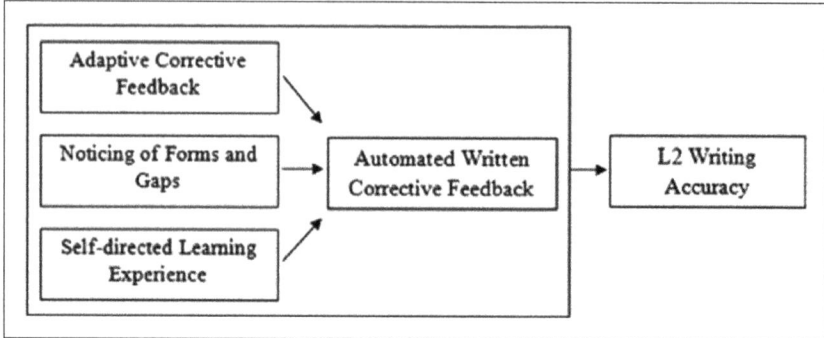

Figure 3.1 AWCF model (in Barrot, 2021, Long, 1996)

the use of Grammarly had a positive effect on both FLA and LA. Learners had positive perceptions toward Grammarly and its help in English writing.

After characterizing eight categories that distinguish AI tools and applications, considering AI technologies' functionalities and ways of use in the language classroom, we will proceed with characterizing AI-powered systems and applications in English language teaching (ELT). Considering scholars' scientific findings, we will pay special attention to the benefits and challenges of their use. Today, it is known that AI-powered systems and applications are being used to improve speaking, reading, and writing skills. These resources provide new ways of teaching and support learners in managing their own learning. AI-powered systems and applications can help students practice a foreign language, such as English, outside class. It should be considered that there are not only the benefits of using AI technologies in English language teaching (ELT) but also challenges. So, in this monograph, we consider the postulates of research by Edmett et al. (2023), Lee et al. (2023), Thompson et al. (2018), Kushmar et al. (2022), and Rowe (2022), the arguments "for" and "against" AI use in ELT, reviewing five key areas (speaking, writing, reading, pedagogy, self-regulation) in which AI is being used in ELT, are summarized below (Table 3.2).

Table 3.2 Benefits (affordances) and challenges of using AI-powered systems and applications in ELT

	Benefits (affordances) of using AI-powered systems and applications in ELT
Speaking	• Pronunciation and intonation are important subskills in speaking that significantly influence the clarity and effectiveness of communication in EFL learning; AI-powered systems and programs available online for learners contribute to enhancing pronunciation;
	• AI-powered systems and applications are employed as a language coach;
	• AI-powered systems and applications are used as coaching systems, employing multiple ways to present information, such as text, images, audio, and video. The learners that use an AI coach are provided with the opportunity to be equipped with the ability to speak more fluently using consistently accurate language structures;
	AI-powered systems and applications are used for speech recognition, adaptive learning, automatic speech analysis, and voice assistance. It helps learners to practice, record, and react to learners pronouncing words, resulting in longer retention of the vocabulary and significant benefits in learning consonant and vowel sound
Writing	• Neural machine translation programs resulted in learners' vocabulary improvement, especially when specialized and unambiguous expressions were involved
	• AI is used for improving writing text and as an AI grammar checker (e.g., the AI-powered tool Grammarly)
	• AI tools provide feedback via spelling and grammar checkers (e.g., the AI-powered tool Grammarly)
Reading	• AI is used for developing the receptive skill (reading and listening)
Pedagogy	• For the pedagogical sphere, the benefit of AI is closely linked to the methods, strategies, and pedagogical techniques used to facilitate ELT. AI-powered systems and applications can help the teacher search for lecture materials. Applying teaching approaches such as a personalized learning approach and a learner-generated-context-based (LGC) approach (Lee et al., 2023) contributes to improving learners' knowledge. Remarkably, the application of AI-powered resources contributes to enhancing learners' vocabulary and improving language fluency and grammatical precision. The LGC AI-powered pedagogical approach also fosters learners' self-autonomous learning ability

Benefits (affordances) of using AI-powered systems and applications in ELT	
Self-regulation	• Using AI promotes self-regulation. Self-regulation means the ability to manage and control one's thoughts, emotions, behaviors, and physiological responses to achieve personal goals and maintain well-being
Challenges using AI-powered systems and applications in ELT	
Technology breakdown	• Technology breakdown is a challenge that implies technical malfunctions and poor connectivity. One specific technology breakdown is incorrect answers from the AI-powered system or application.
Limited capabilities	• We discuss limited capabilities when learners require more advanced functionality from the AI-powered system or application. For instance, some learners want better chatbot capacity, while others wish for more "natural" interactions (Thompson et al., 2018). Limited capabilities can make learners uninterested in using the chatbot or another AI tool or application.
Fear	• Fear takes several forms, including 1) a lack of clarity on how personal information would be stored and shared, 2) fear of the unknown, i.e., uncertainty about how the AI is operating, and 3) fear of losing a natural learning environment and, along with it, real emotions, anxiety for instance, connected to learning (Kushmar et al., 2022).
Standard-izing languages and ideologies	• Standardizing languages and ideologies occurs as one of the most compelling challenges. Let us provide an example: "Rowe's (2022) study of learners in a second-grade American classroom found that Google Translate's programming appeared to carry messages about what is considered appropriate and standard language use, disregarding nuances in language groups. One learner using the tool found that Tagalog was not listed as a language by Google Translate, and the only available option for the Tagalog-speaking pupil when translating her own language to English was Filipino (which has been the official standardised language of the Philippines since 1987). Rowe (2022, p.884) reports that this left the learner 'in essence, engaged in a negotiation of what counts as a language, who decides what it is called, and which language was "correct."' This suggests that by recognizing some historical and political language boundaries over others, Google might re-enforce standardised language use" (in Edmett et al., 2023).

Researchers Helen Crompton, Adam Edmett, Neenaz Ichaporia, and Diane Burke, in a recently published paper, applied the grounded coding approach. The use of this approach can be considered very pertinent, as its application allowed the authors to reveal skills related to reading, speaking, and writing. Moreover, this team of researchers (Crompton et al., 2024) created axial codes for speaking, writing, reading, pedagogy, self-regulation, and challenges (Figures 3.2–3.7).

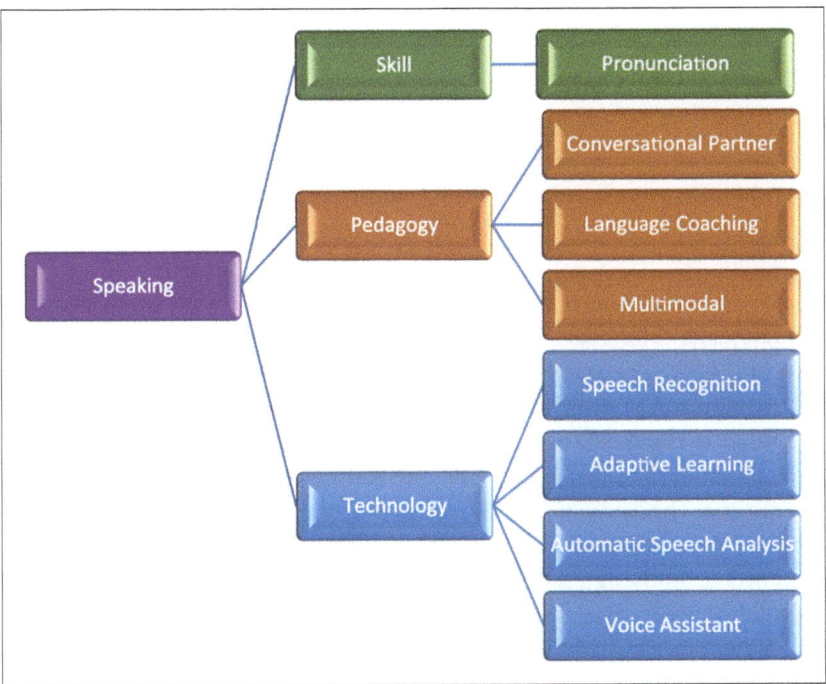

Figure 3.2 Axial codes for speaking (Source: Crompton et al., 2024)

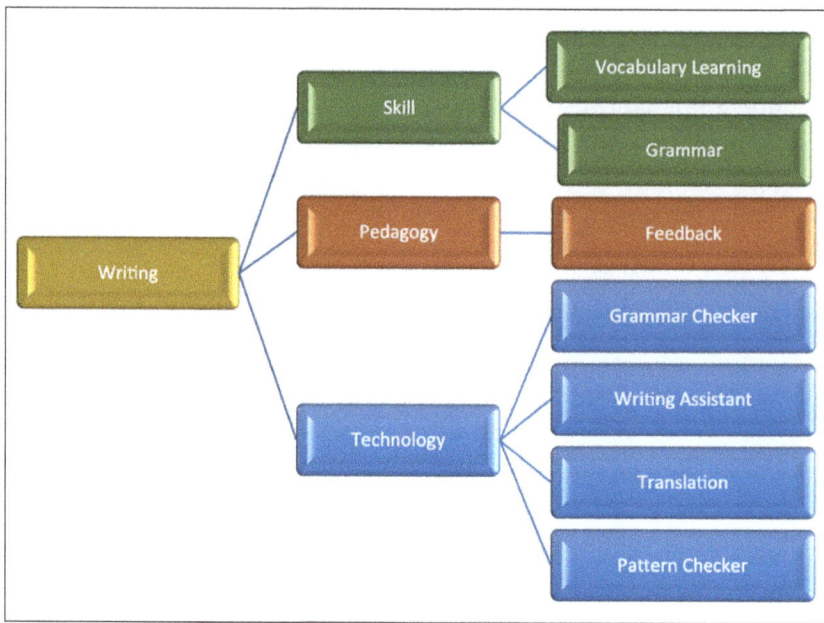

Figure 3.3 Axial codes for writing (Source: Crompton et al., 2024)

AI, viewed in the context of teaching English speaking, is a collateral resource as a language tutor/coach (Speechify, Leya, and Loora) or a conversational partner (ELSA). Practicing speaking skills with AI resources stands on three pillars (Figure 3.2): skill, pedagogy, and technology.

Similar to the speaking code (Figure 3.2), axis codes for writing stand on three pillars such as pedagogy, skill, and technology (Figure 3.3). Exploiting AI grammar checkers, such as Grammarly, a spelling and grammar checker, provide real-time correction feedback, which undoubtedly contributes significantly to studying English.

Notably, vocabulary is the only subskill in axial codes for reading (Figure 3.4). AI, in our view, adds a game element to the process of learning to read in English. This game element is in text-to-speech technology—Speechify (Figure 3.8), as an example.

AI TECHNOLOGY AND LANGUAGE EDUCATION

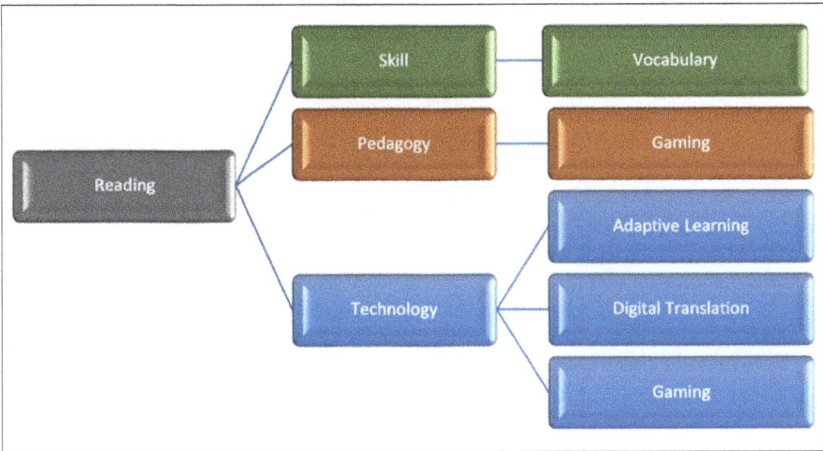

Figure 3.4 Axial codes for reading (Source: Crompton et al., 2024)

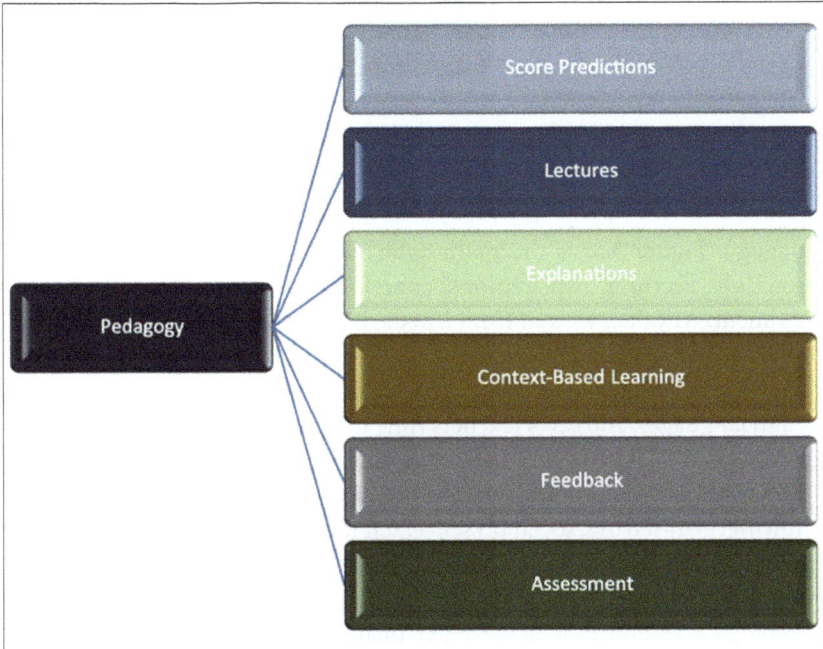

Figure 3.5 Axial codes for pedagogy (Source: Crompton et al., 2024)

149

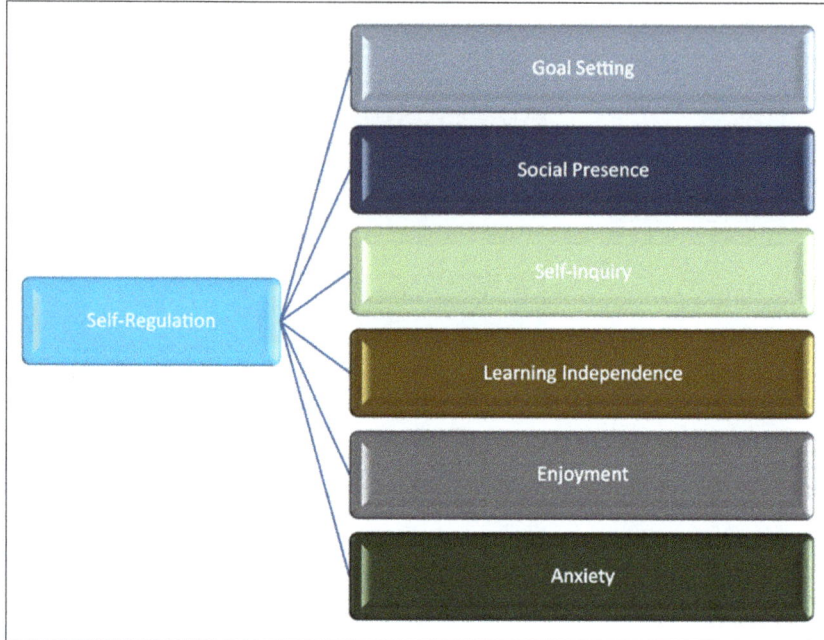

Figure 3.6 Axial codes for self-regulation (Source: Crompton et al., 2024)

Pedagogy (Figure 3.5) is the main code. There are six axial codes: score predictions, lectures, explanations, context-based learning, feedback, and assessment. AI value, viewed from a pedagogical point of view, expands the boundaries for teaching in the time of AI we live in now. AI allows the creation of learner-centered content, modernizes traditional forms of teaching foreign languages, and saves teachers time in preparing exercises for lessons. An example is "CATHOVEN Language Hub," an AI time-saver for teachers.

Self-regulation (Figure 3.6) stands on six axial codes. The axial codes are the following: goal-setting, social presence, self-inquiry, learning independence, enjoyment, and anxiety. It is essential to remember that "self-regulation refers to the ability to manage and control one's thoughts, emotions, behaviours and physiological responses to achieve personal goals and maintain well-being. Affect can influence the choices and actions the pupil takes. For example,

AI TECHNOLOGY AND LANGUAGE EDUCATION

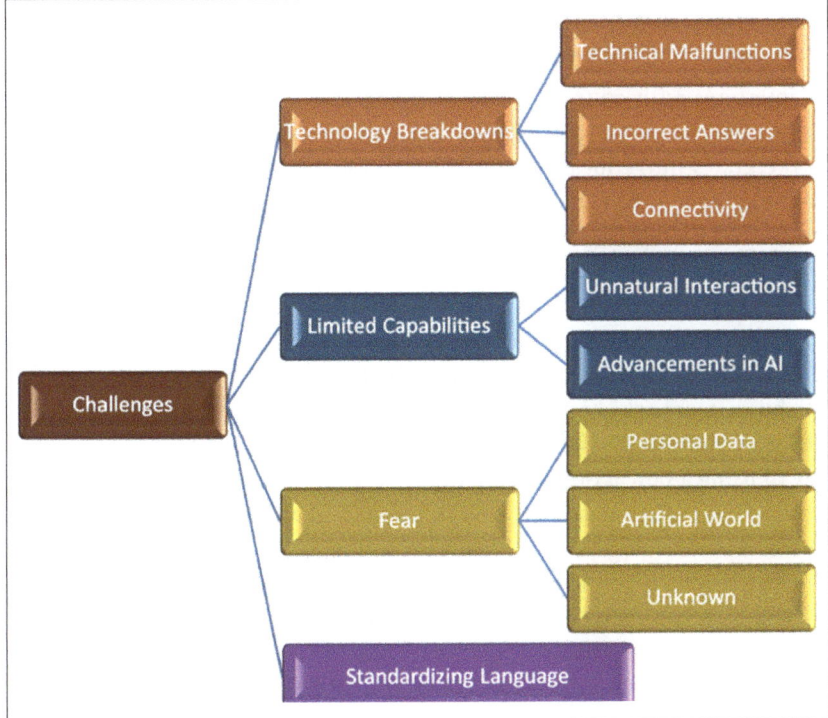

Figure 3.7 Challenges with the use of AI in ELT/L (Source: Crompton et al., 2024)

someone in a negative mood might be more prone to impulsive decisions or avoidance behaviours" (Crompton et al., 2024). Also, the important aspect is that the exploit of AI-powered tools, chatbots, for instance, to promote self-regulation could aid learners in becoming more autonomous, lessening anxiety and concerns about speaking in a foreign language, particularly English.

Four axial codes for challenges (Figure 3.7) are technology breakdown, fear, limited capabilities, and standardizing language. Limited capability is an axial code that refers to people "wanting the AI systems to do more of what the system can do in a more advanced way" (Crompton et al., 2024). Fear is another axial code that consists of three concerns related to AI use: (1) the leaking of personal data, (2) fear of the unknown, and (3) fear of losing real emotions and sense of reality.

3.2. The path of integration AI in the language classroom: practical dimensions

AI interferes with education as youngsters spend much time online. This chapter delineates the valuable practical aspects of AI for language education. AI integrates into education, and AI technologies impact the learning and teaching process. AI can effectively ease teachers' work. Pedagogues can use innovative, disruptive solutions to deliver AI-supported language teaching. AI helps optimize the learning background, offering a highly personalized approach to teaching.

Traditionally, language education (Tajeddin and Griffiths, 2023) is characterized by applying teacher-centered instruction, standardized curricula, syllabuses, textbooks, tests, teachers' guides, and other educational materials. However, the use of AI brings new ideas and approaches to education. In this way of using a new AI-powered paradigm, an innovative teaching–learning environment is being formed, and solutions and applications are being generated. AI technologies offer creative solutions to meet the evolving needs of learners.

NLP directs researchers' and pedagogues' attention to the branch of science that implies new technological advances, to be more concrete—AI. NLP (Khurana et al., 2023; Abro et al., 2023) is a subfield of AI that uses machine learning to assist computers in communicating with people. NLP algorithms can assess grammatical accuracy and correct vocabulary, providing learners punctual feedback and corrective guidance.

Gamified AI-powered language-learning apps (e.g., Duolingo, Babbel, Rosetta Stone) have and use game mechanics to motivate learners to learn and promote active participation. VR and AR technologies enable learners to explore the virtual "world" and engage in simulated discussions and talks. Through this "speaking practice," learners develop language skills meaningfully.

To learn to speak—the task of learning a new language. AI offers resources that have been developed to achieve excellence in speaking. Nowadays, AI apps provide the opportunity to practice a foreign language realistically with a "human-like" conversation tutor or just with an interlocutor who knows how to maintain a conversation on any topic and at any level and help the

AI TECHNOLOGY AND LANGUAGE EDUCATION

learner deal with errors. AI apps are different. We have come across AI apps so far: Speechify, ELSA, Leya, and Loora.

A view on Speechify from a practical standpoint

"Speechify" (Figure 3.8) exploits text-to-speech technology and converts written content. Learners can listen to the content. "Speechify" is accessible on the internet. "Speechify" is designed to cater to individuals with learning disabilities, visual impairments, and learners who prefer auditory learning. This AI-powered platform provides a spoken version of the written content and ensures equal access for learners to information.

"Speechify" is a valuable learning tool for students, teachers, and employees. By converting written content to audio, "Speechify" facilitates comprehension of the foreign language and eases memorizing.

To register in "Speechify," the learner is asked to answer some questions (Figure 3.9). The questions relate to the user's reading preferences and the result the user wants to achieve while relying on "Speechify" AI power.

"Speechify" is an AI platform that assists teachers of foreign languages. Here are some characteristics of "Speechify":

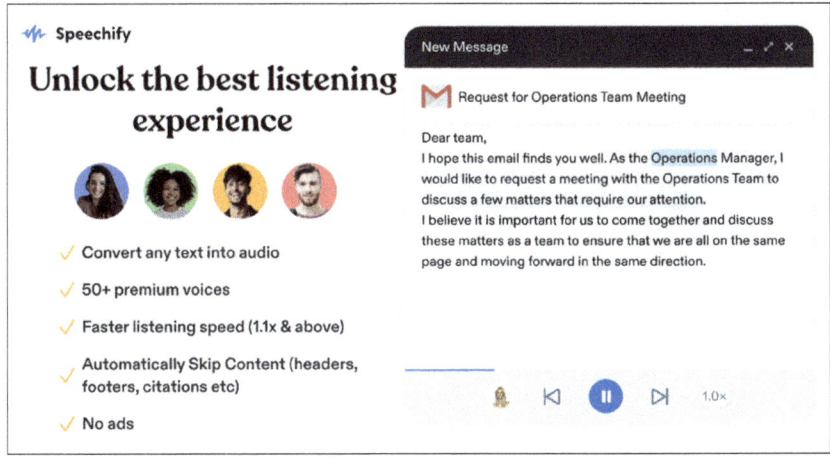

Figure 3.8 Speechify <www.speechify.com>: key characteristics

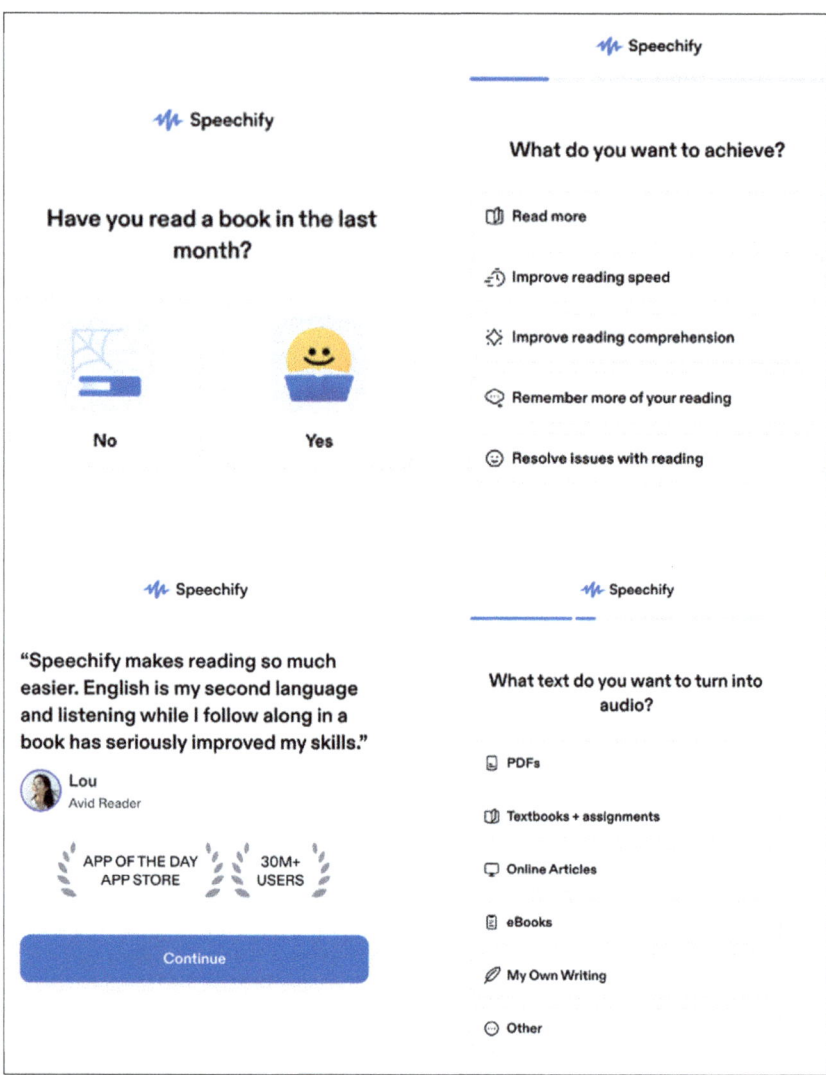

Figure 3.9 Speechify: questions for learners

- Presents text-to-speech technology (it converts written content into spoken words and phrases); the learner can choose the voice of the speaker who will read the text (Figure 3.10);
- Eases personalized learning (it tailors the content of education to the individual preferences and learners' needs);
- Represents ITS (it is an intelligent tutoring system that fosters critical thinking and helps develop problem-solving skills);
- Creates a collaborative learning environment (it allows collaborative learning, facilitates communication, and eases knowledge sharing).

Thus, "Speechify" is an AI app that allows the creation of a "speaking" learning location.

A view on ELSA from a practical standpoint

ELSA (English Language Speech Assistant) is an AI-powered English-speaking coach platform (Figure 3.11). Users can speak English in short dialogues and get instant feedback from ELSA's proprietary AI technology.

It helps achieve great strides in mastering the English language. The use of ELSA enhances pronunciation, fluency, and comprehension. Here are some characteristics of ELSA:

- It contributes to improving pronunciation (it utilizes speech recognition technology for analyzing the learner's pronunciation; it provides real-time feedback; it compares the learner's speech patterns with the native speaker's accents and then offers suggestions for improving pronunciation accuracy, intonation, and rhythm);
- It supports ideas of personalized learning (it takes into consideration each learner's level of proficiency);
- It represents ITS (it is an intelligent tutoring system that fosters critical thinking and helps develop problem-solving skills);
- It is accessible on the Internet.

After pronouncing the word, ELSA provides real-time feedback (Figure 3.12):
After pronouncing the sentence, ELSA provides real-time feedback (Figure 3.13):

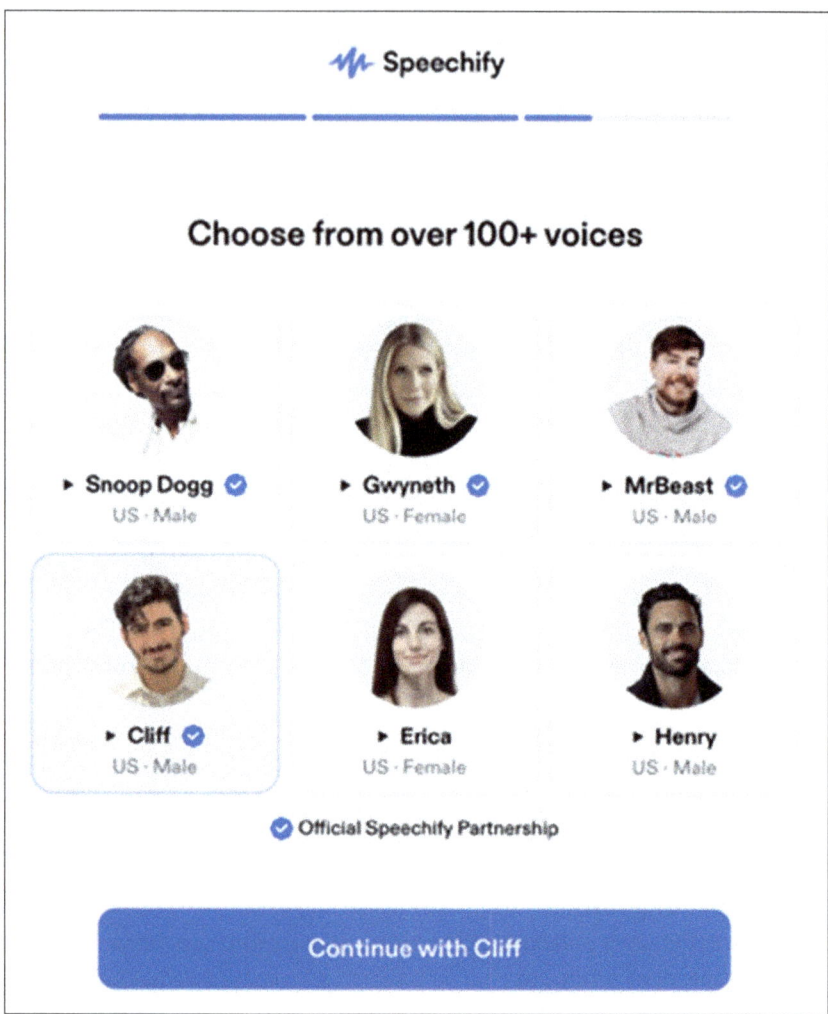

Figure 3.10 Speechify: the possibility to choose the speaker by selecting the voice that the user likes

AI TECHNOLOGY AND LANGUAGE EDUCATION

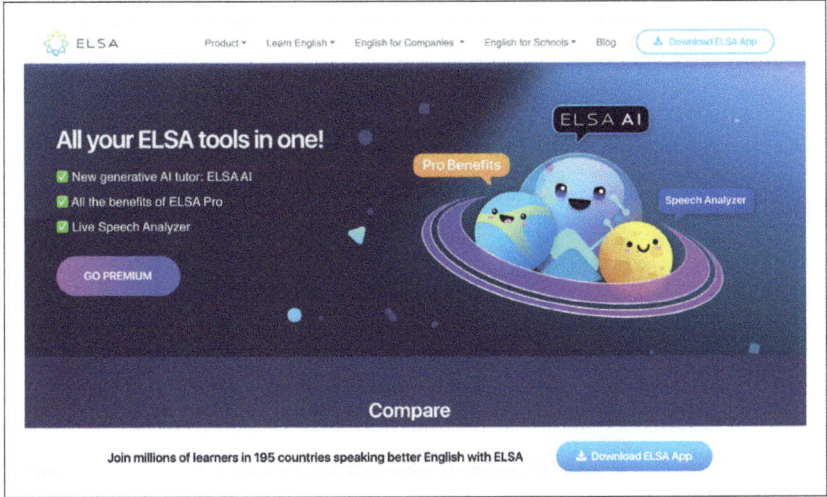

Figure 3.11 ELSA <www.elsaspeak.com>: possibilities for learners

The conversation with ELSA also contributes to improving speech skills. ELSA asks the questions, then the learner reads the proposed answer. Real-time feedback provides an evaluation of the pronounced information (Figure 3.14:)

Thus, ELSA is an AI platform. This AI platform helps improve the pronunciation of English words, sentences, and conversation practice. ELSA provides the following features: voice-enabled role-play, realistic speaking practice, "create your own scenario" feature, plenty of mini-lessons on various skills and topics, daily recommended practice, progress tracking and rewards, access to certificate courses (IELTS, Oxford, HarperCollins, EIKEN, Pearson PTE), access to the US Citizenship Test Preparation Course, AI-powered feedback on fluency, grammar and vocabulary, guided practice for interviews, presentations and exams, dynamic transcript, paraphrasing feature and vocabulary recommendations.

A view on "Leya" from a practical standpoint

Leya (<leyaai.com>) is an AI platform. It is, let's say, a personal AI tutor of English. It engages in conversation with an AI tutor. Touting their product,

157

Figure 3.12 ELSA: pronunciation of the word and real-time feedback

Figure 3.13 ELSA: pronunciation of the sentence and real-time feedback

CHAPTER 3

Figure 3.14 Communication with ELSA with real-time feedback

the creators of the "Leya" website made a table to compare "Leya's" options with a human tutor (Table 3.3).

There are benefits to using Leya's AI English tutor (Figure 3.15). They are compelling for users of the "Leya" website. Let's list them: the accessibility of AI-powered activities, receiving immediate feedback, and accessibility to various learning topics.

By leveraging "Leya," the English language learner can improve listening, grammar, reading, vocabulary, writing, and speaking skills.

A view on "Loora.ai" from a practical standpoint

Loora (loora.ai) is an AI English tutor. It allows practicing real-life English in real-life situations, such as business meetings, job interviews, friendly chats about the latest book you have read, the show you could see, and discussing business and fashion-related topics. It gives real-time feedback and helps perfect pronunciation, fix grammar, embrace rephrasing, and enhance accent.

Loora provides services for business teams. Daily lessons contribute to gradual improvement in speaking English fluently (Figure 3.16).

Based on the analysis, the weaknesses and strengths of each AI app (Speechify, ELSA, Leya, and Loora) can be specified (Table 3.4).

Table 3.3 A comparison table of options provided by Leya, an AI English tutor, with human tutor work

Leya—an AI English tutor	A human tutor
Leya sounds and speaks like a real tutor	A person
No need to plan ahead	A tutor has a busy schedule; the learner has to find a time that works for both
Leya is available all day, every day	The learner pays for each tutoring hour
Leya provides immediate feedback on the learner's conversational abilities	A tutor follows traditional lessons of 45–60 minutes, which are not flexible
When speaking to Leya, the learner does not feel pressured or anxious. AI tutor does not judge	The learner can initially feel uncomfortable when having the lesson with a tutor

Source: <leyaai.com>

Figure 3.15 Leya—AI English tutor: compelling benefits for users (Source: <leyaai.com>)

Figure 3.16 Loora—an AI English tutor: service for the business team (Source: loora.ai)

A view on the "CATHOVEN Language Hub" from a practical standpoint

"CATHOVEN Language Hub" (<cathoven.com>) is a time-saver for teachers who need quick but robust assessment or to create tasks for the upcoming lesson. For the lesson, the CATHOVEN Language Hub can create readings with specific vocabulary and grammar, create listening exercises from a

Table 3.4 Weaknesses and strengths of AI apps

Strengths	Weaknesses
AI apps provide real-time feedback (Loora, Leya), indicate pronunciation mistakes (ELSA)	All apps can occasionally give incorrect advice on vocabulary or grammar
The learner has an "interviewer" and can even choose the voice the learner likes (Speechify)	The voices of AI apps are computer-generated. They are different from voices that we hear in the streets of London or New York, for example, or encounter in real situations
AI apps offer conversations with different (according to CEFR) levels of vocabulary (but it depends on an AI app) and provide the possibility to practice real-life English in real-life situations (Loora)	Chatting lacks human connection. Not for all levels
Easy access (via the Internet)	Paid access to certain content

video, rewrite the text in a different language level, assess ESL/EFL language difficulty of the text, provide feedback on students' writing, find vocabulary and grammar in a video, create comprehensive exercises from the text (for example, multiple-choice task: Figure 3.17), check the text's reading level.

Creating comprehension exercises (e.g., multiple-choice tasks, true/false, short answer questions, and true/false/not given) from the text is quite a feasible task for the CATHOVEN Language Hub.

Teachers can download the created tasks in PDF format or Docx format or copy the task (Figure 3.18):

A True/False/Not given exercise generated by CATHOVEN Language Hub is presented in Figure 3.19:

Providing feedback on students' writing is one of the teacher's tasks. CATHOVEN Language Hub can do it (Figure 3.20):

AI can give the affordances and bring challenges. Chapter 3 of the monograph is unique in three points: (1) the analysis of scientific publications, compelling new theories on foreign language learning and teaching, is carried out; (2) an up-to-date review of AI apps that are helpful in the training correct pronunciation and skills of free language speaking, in writing without errors,

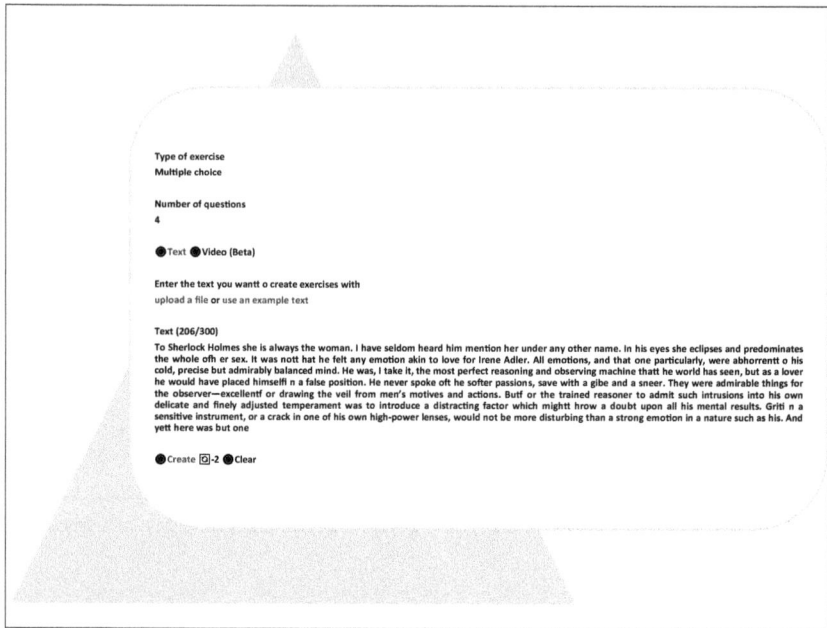

Figure 3.17 CATHOVEN Language Hub: a multiple-choice task (Analysis by <nexthub.cathoven.com>); Appendix A—questions/answers.

and reading is offered; (3) an examination of the time-saver for language teachers—CATHOVEN Language Hub—is done. The chapter's data described five areas in which AI can accompany ELT. These areas are self-regulation, pedagogy, reading, writing, and speaking. Each of these areas is linked with axial codes. The challenges were also delineated. The monograph chapter attracts attention to the need to explain the mechanism of AI application in language education, clarifying what AI is and how to exploit the benefits of AI apps and AI teachers' time-savers.

3.3. Recommendations for the use of AI technology in language education of students of economics

AI technology is a new stage in language education. AI technology for language education has been a hotly debated and top research topic. A lot of interesting and fruitful scientific publications (Table 3.1) have enabled both

Download exercise
Choose the sections you want to download:
☑ Text
☑ Questions
☑ Answers
☐ Explanation
☑ Answer location

Display answers and other extra information at the end of
☐ Each question
☑ Page

Preview (select the questions you want to include):
Read the passage and choose the correct answer.
To Sherlock Holmes she is always the woman. I have seldom heard him mention her under any other name. In his eyes she eclipses and predominates the whole of her sex. It was not that he felt any emotion akin to love for Irene Adler. All emotions, and that one particularly, were abhorrent to his cold, precise but admirably balanced mind. He was, I take it, the most perfect reasoning and observing machine that the world has seen, but as a lover he would have placed himself in a false position. He never spoke of the softer passions, save with a gibe and a sneer. They were admirable things for the observer—excellent for drawing the veil from men's motives and actions. But for the trained reasoner to admit such intrusions into his own delicate and finely adjusted temperament was to introduce a distracting factor which might throw a doubt upon all his mental results. Grit in a sensitive instrument, or a crack in one of his own high-power lenses, would not be more disturbing than a strong emotion in a nature such as his. And yet there was but one woman to him, and that woman was Irene Adler, of dubious and questionable memory.

☑1. What is the primary reason Sherlock Holmes admires Irene Adler?
☐ A. She is unique and dominates over others in his perception.
☐ B. He has a romantic affection for her.
☐ C. Her manner is always a source of emotional comfort for him.
☐ D. She is a representation of all women in his life.

☑2. How does Sherlock Holmes view emotions in relation to his work as a detective? (…)
● Copy ● Download as PDF ● Download as Doc

Download exercise
Choose the sections you want to download:
☑ Text
☑ Questions
☑ Answers
☐ Explanation
☑ Answer location

Display answers and other extra information as:
☐ Each question
☑ A page

Preview (select the questions you want to include):
1.
Answer: A.
Answer Location: "In his eyes she eclipses and predominates the whole of her sex."

2.
Answer: D.
Answer Location: "But for the trained reasoner to admit such intrusions into his own delicate and finely adjusted temperament was to introduce a distracting factor which might throw a doubt upon all his mental results."

3.
Answer: C.
Answer Location: "All emotions, and that one particularly, were abhorrent to his cold, precise but admirably balanced mind."

4.
Answer: A.
Answer Location: "He was, I take it, the most perfect reasoning and observing machine that the world has seen."

● Copy ● Download as PDF ● Download as Doc

Figure 3.18 CATHOVEN Language Hub: the download options and the option for copying a multiple-choice task (analysis by <nexthub.cathoven.com>)

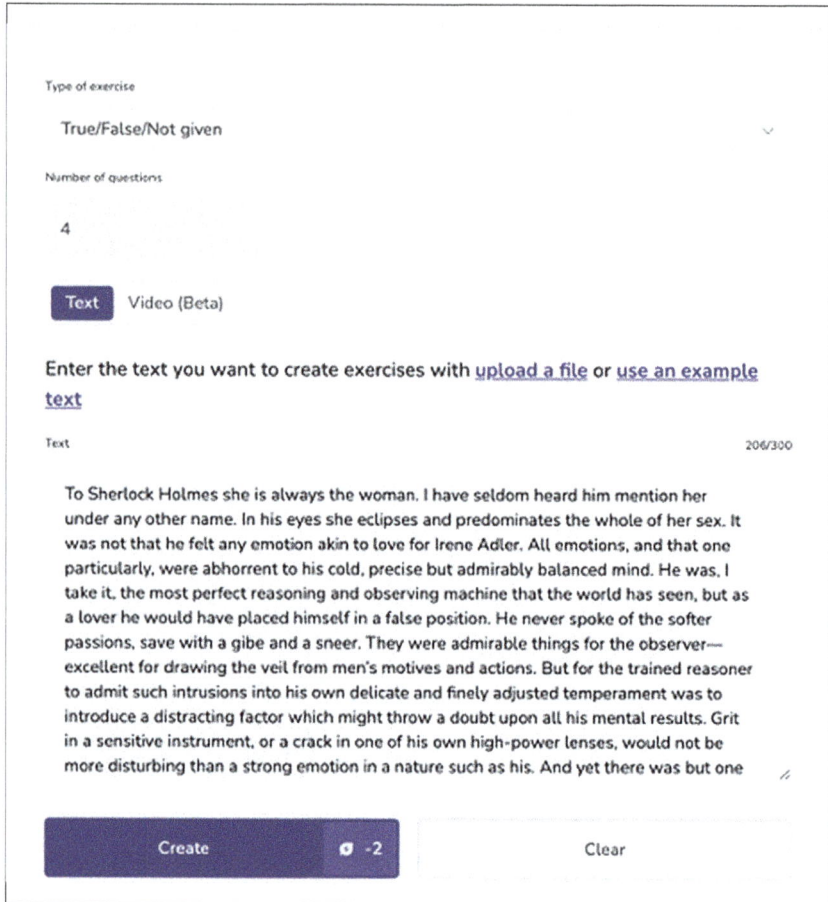

Figure 3.19 CATHOVEN Language Hub: A True/False/Not given exercise (Analysis by <nexthub.cathoven.com>); Appendix B—questions

language teaching professionals and scholars to discover more about the application of AI in language education. AI technology is moving rapidly, so ongoing research is necessary to comprehend better how students view and think about language education. For this, we offered a questionnaire to students majoring in economics who are studying at the University of Economics in Bratislava. The questionnaire comprised open questions and

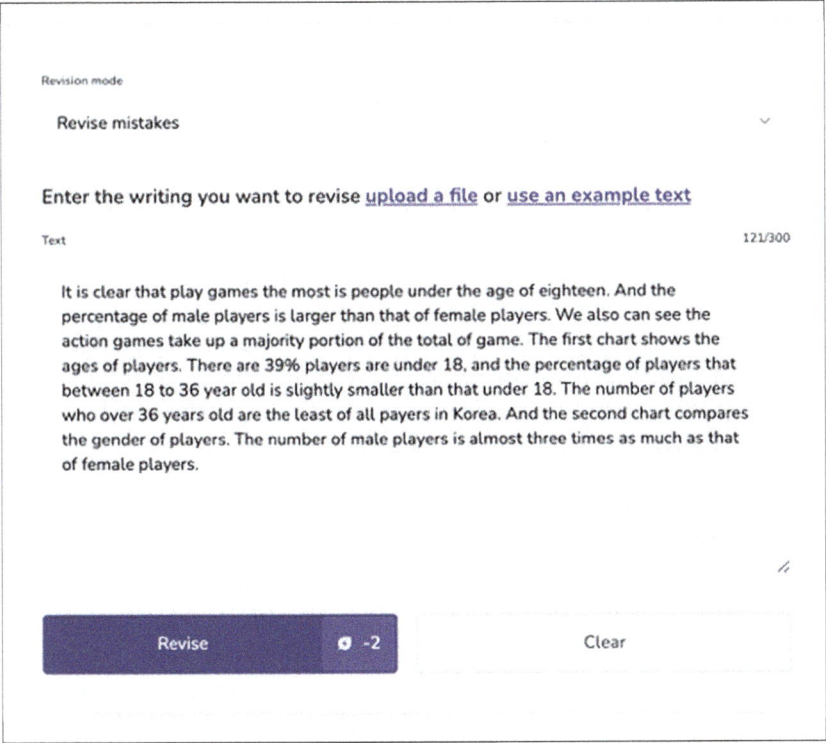

Figure 3.20 CATHOVEN Language Hub: Providing feedback on students' writing (Analysis by <nexthub.cathoven.com>); Appendix C—feedback

Likert-scale questions to examine the opinions of learners about teaching languages at the university. Participants' opinions were collected. The purpose of the questionnaire was to analyze the views of students; then, based on the responses received, propose recommendations for the use of AI technology in language education of students of economics.

3.3.1 Questionnaire analysis and synthesis

The survey participants, 338 students, are bachelor (91.4%) and master (8.6%) students. The survey (Appendix D) was conducted between October 16, 2023, and December 01, 2023. The survey participants study at different faculties of the University of Economics in Bratislava: Faculty of Commerce (31.1%),

Faculty of Economics and Finance (19.4%), Faculty of Economic Informatics (12.9), Faculty of International Relations (9.5%), Faculty of Applied Languages (27.1%). According to a survey carried out, the overwhelming majority of respondents learn English as a first foreign language (311 students). Survey participants also learned German as their first foreign language (7 students), Spanish (5 students), and Slovak (5 students). At university, students learn a second foreign language. As the survey shows, 107 students learned German as a second foreign language, 55 students learned Spanish as a second foreign language, 19 students learned French as a second foreign language that they learned at the university, 16 Russian, and 3 Italian. Students also learn a third foreign language. As the survey shows, 12 students learn Spanish as a third foreign language, and three students learn French as a third foreign language; for three students, Italian is the third foreign language.

What are the top reasons why students of economics learn foreign languages?

The method of analysis and synthesis (Riemann, 1900, 1984, 1990, 1995, in Ritchey, 1991, in Bell et al., 2022) was employed in this chapter of the monograph. As Tom Ritchey (1991) wrote: "Analysis is defined as the procedure by which we break down an intellectual or substantial whole into parts or components. Synthesis is defined as the opposite procedure: to combine separate elements or components in order to form a coherent whole." Further, explaining the essence of the method of analysis and synthesis, Ritchey (1991) continued: "Analysis and synthesis, as scientific methods, always go hand in hand; they complement one another. Every synthesis is built upon the results of a preceding analysis, and every analysis requires a subsequent synthesis in order to verify and correct its results. In this context, to regard one method as being inherently better than the other is meaningless." For the current study, the application of the method of analysis and synthesis is essential as the application allows us to seek the views of survey participants on reasons that have led to choosing to learn English as a first foreign language: German, Spanish, French, Russian and Italian—the second foreign language to learn; and the third foreign language for learning it. Then, it enables forming a coherent picture of the points in favor of learning foreign languages. Based on the data (questionnaire analysis for determining reasons why students of economics learn English as the first foreign language at Bratislava University of Economics and Business; the opinions of 311 students out of 338 survey participants were analyzed; every survey participant could write from one to three reasons that, in their view, are essential) of the conducted analysis, using the method of analysis and synthesis (Figure 3.21), the reasons why students of economics learn English as a first foreign language were determined (Figure 3.22): (1) English is considered a global language; (2) English opens up career opportunities; (3) English is a mandatory subject; (4) English is easy to learn; (5) for personal improvement; (6) the practical application of English in everyday life; real-world relevance; (7) cultural benefits.

Based on the data (questionnaire analysis for determining reasons why *students of economics* learn the *second foreign language* (German, Spanish, French, Russian or Italian) at Bratislava University of Economics and Business; the opinions of 213 students out of 338 survey participants were analyzed; every survey participant could write from one to three reasons that, in their view, are essential) of the conducted analysis, using the method of analysis and synthesis (Figure 3.21), the reasons why students of economics learn the second foreign language (German, Spanish, French, Russian or Italian) were determined (Figure 3.23): (1) career opportunities: (a) a good working knowledge of the second language opens up career opportunities, (b) the geographical relevance (many survey participants hold the opinion that German is widely spoken in countries neighboring Slovakia; a lot of German-speaking companies operate in Slovakia), (c) the economic significance (French, Spanish and German are viewed as languages of international business and trade); (2) personal development: (a) one of the best ways to expand skills, (b) the educational requirement (the second language is a mandatory subject in school); (3) an interest in travels and other cultures; (4) the language is easy to learn; (5) the professional aspirations: (a) students of economics perceive the knowledge of the second language as a tool to access economic literature or for research purposes; (b) French and Russian are seen as an important tool in the diplomatic sphere.

Based on the data (questionnaire analysis for determining reasons why *students of economics* learn the *third foreign language* (Spanish, French, or Italian) at Bratislava University of Economics and Business; the opinions of 54 students out of 338 survey participants were analyzed; every survey participant could

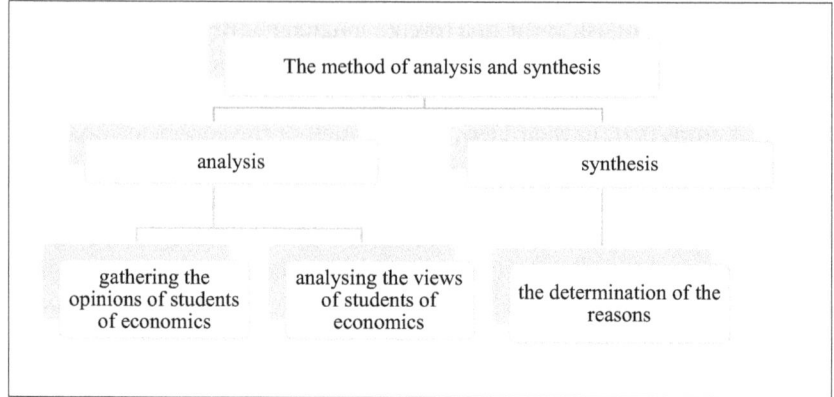

Figure 3.21 The use of the method of analysis and synthesis for the determination of the reasons why students of economics learn the first foreign language

write from one to three reasons that, in their view, are essential)of the conducted analysis, using the method of analysis and synthesis (Figure 3.21), the reasons why students of economics learn a third foreign language (Spanish, French, or Italian) were determined (Figure 3.24): (1) career opportunities; (2) personal interests; (3) the language is easy to learn; (4) an interest in travels; (5) educational requirement (within the university curriculum).

The importance of learning a foreign language within the context of general education (the assessment of the significance in the order of importance by way of a five-point Likert scale from 1 [extremely important] to 5 [not at all important])

Most surveyed participants believe learning a foreign language is extremely important (145 students/43.3%). Many students think it is very important (77/23%). It is moderately important for 51 (15.2%) of survey participants. It is slightly important for 34 (10.1%) students. It is not at all important for 28 (8.4%) students.

The importance of learning a foreign language from the point of view of professional use after graduation (the assessment of the significance in the order of importance by way of a five-point Likert scale from 1 [extremely important] to 5 [not at all important])

AI TECHNOLOGY AND LANGUAGE EDUCATION

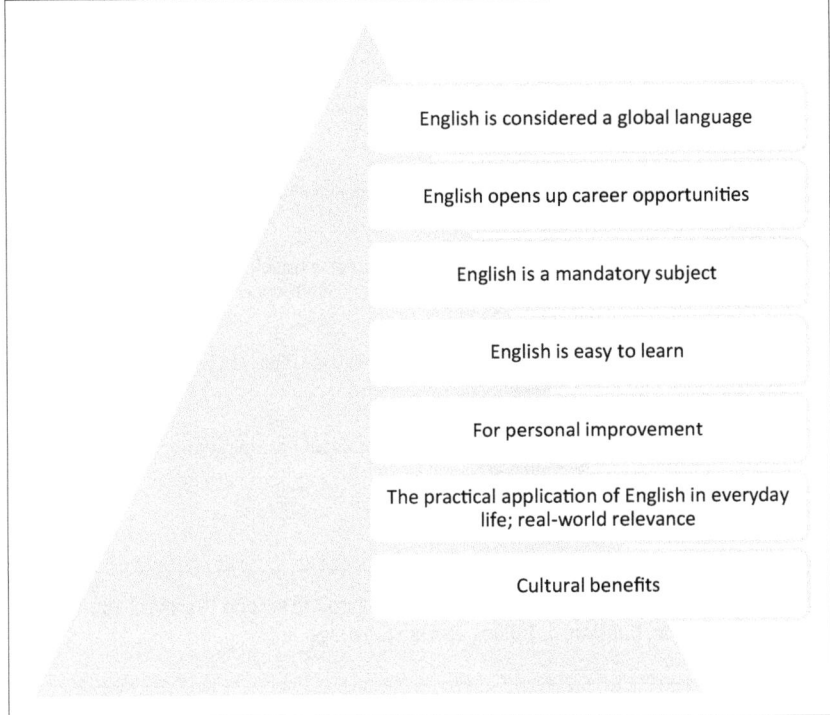

Figure 3.22 Reasons why students of economics learn English as the first foreign language

Most surveyed participants believe that knowledge of foreign languages will be extremely important (153 students/45.5%) in future professional activity. Many students think it is very important (75/22.6%). It is moderately important for 44 (13.1%) of survey participants. It is slightly important for 30 (8.9%) students. It is not at all important for 33 (9.8%) students.

Self-assessment of knowledge of the foreign languages

The first foreign language
The survey participants assessed their knowledge of the first foreign language using a five-point Likert scale from 1 (excellent knowledge) to 5 (poor

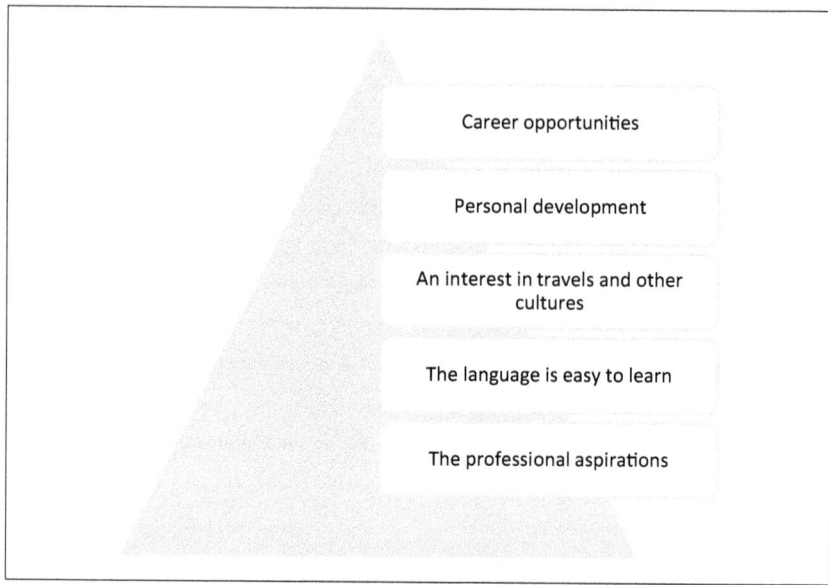

Figure 3.23 Reasons why students of economics learn the second (German, Spanish, French, Russian, or Italian) foreign language

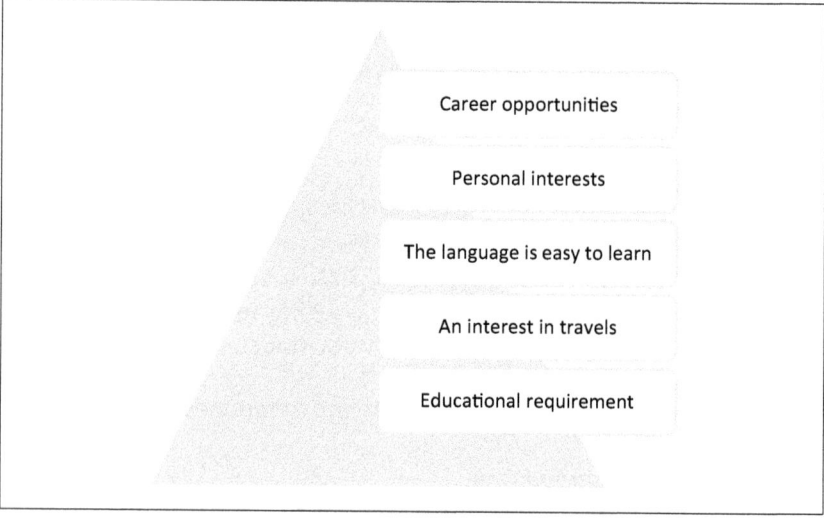

Figure 3.24 Reasons why students of economics learn a third (Spanish, French, or Italian) foreign language

knowledge). Many surveyed participants assessed their knowledge as excellent (96 students/28.7%). The vast majority (112/33.5%) of the respondents think their knowledge is good. Also, many respondents (75/22.5%) believe their level of knowledge is average. Some students (33/9.9%) think their knowledge is not so good. And some students (18/5.4%) think their knowledge is very poor.

The second foreign language

The survey participants assessed their knowledge of the second foreign language using a five-point Likert scale from 1 (excellent knowledge) to 5 (poor knowledge). The vast majority of the surveyed participants assessed their knowledge as excellent (67 students/34.7%). Many (42/21.8%) respondents think their knowledge is good. Also, many respondents (53/27.5%) believe their level of knowledge is average. Some students (18/9.3%) think their knowledge is not so good. And some students (13/6.7%) think their knowledge is very poor.

The third foreign language

The survey participants assessed knowledge of a third foreign language using a five-point Likert scale from 1 (excellent knowledge) to 5 (poor knowledge). The vast majority of the surveyed participants assessed their knowledge as excellent (20 students/31.7%). Some (8/12.7%) respondents think their knowledge is good. Many respondents (19/30.2%) believe their level of knowledge is average. One student thinks that his/her knowledge is not so good. And some students (15/23.8%) think their knowledge is very poor.

Benefits from knowledge of foreign languages and plans for using the knowledge in future professional activity

Based on the data (questionnaire analysis for determining benefits from knowledge of foreign languages and plans for using the knowledge in the professional activity was conducted: views of the students at Bratislava University of Economics and Business were analyzed; 285 students out of 338 survey participants expressed their opinion on the matter) of the conducted analysis, using the method of analysis and synthesis (Figure 3.21), the benefits from knowledge of foreign languages and students' plans for using the knowledge in the professional activity were determined. The benefits are the following (Figure 3.25): (1) increased employability; (2) professional

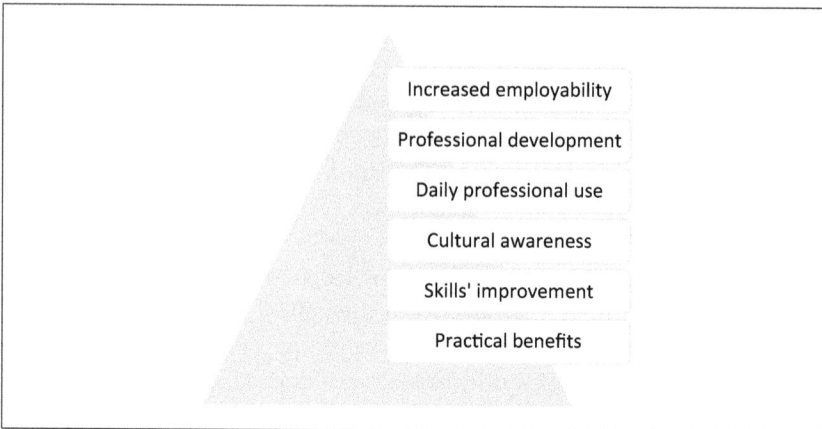

Figure 3.25 Benefits from knowledge of foreign languages: students' views

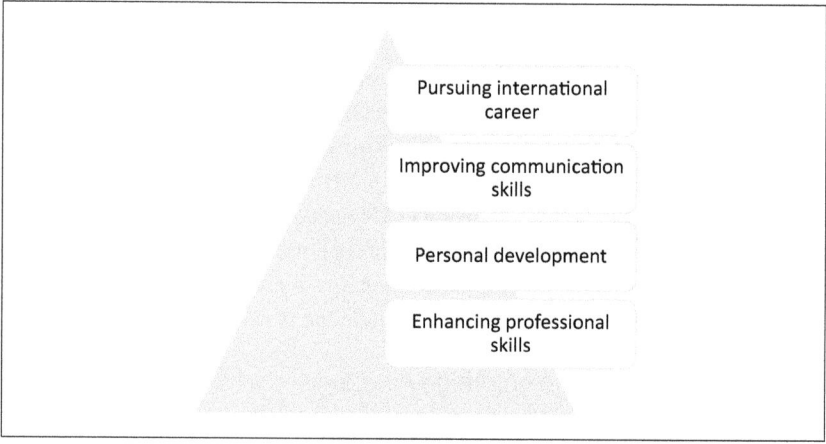

Figure 3.26 Plans for the use of the knowledge of foreign languages in the professional activity

development; (3) daily professional use; (4) cultural awareness; (5) skills improvement; (6) practical benefits. Here are the plans for using the knowledge of foreign languages in professional activity (Figure 3.26): (1) pursuing an international career (the possibility of working abroad; establishing a business that operates globally); (2) improving communication skills (students plan to use their foreign language skills to communicate with clients, collaborate in

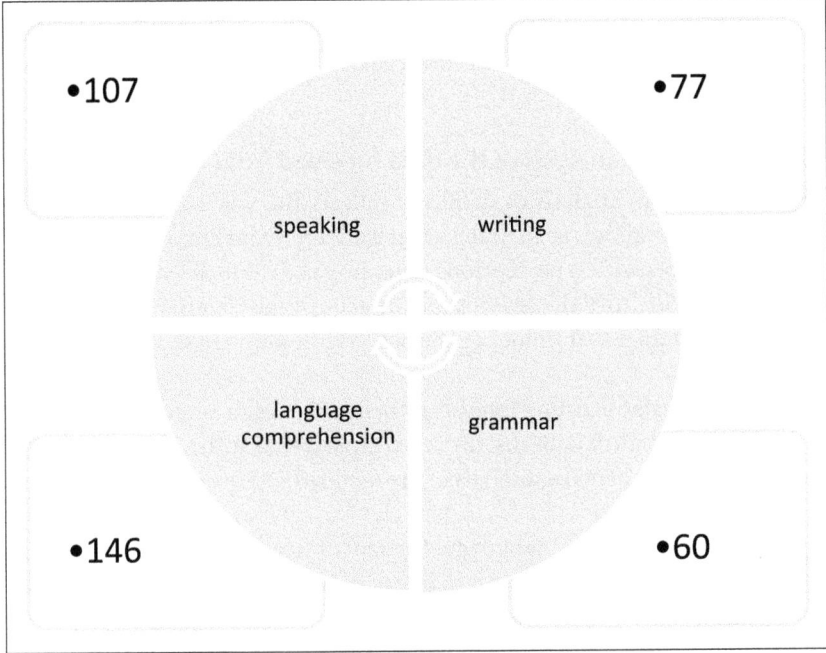

Figure 3.27 The development of speaking and writing skills, grammar competence, and language comprehension in class

international teams, and represent companies abroad; effective negotiation and cross-border communication were mentioned as key professional objectives); (3) personal development; (4) enhancing professional skills.

Skills and competence development and the ability to understand the different elements of spoken and written language

In foreign language classes, teachers focus on the cultivation of linguistic skills (speaking and writing), language competence (grammar), and the ability to understand the different elements of spoken and written language (language comprehension). Which of them are developed to a greater extent, and which are developed to a lesser extent thanks to the foreign language classes? This question was addressed to the survey participants. Students (314 survey participants) set their priorities in this matter (Figure 3.27). Each student could choose more than one of the proposed options. The survey results show that in class, most attention is paid to the development of language comprehension (146 students think so).

3.3.2 Qualitative analysis of the data received from the survey

The questionnaire analysis of data that reflects the views of students who study in Slovakia allows us to understand students' (who major in economics and business specialties) perceptions regarding the value of learning foreign languages and identify the reasons for the acquisition of languages, and understand the factors that influence the perceived value. These are as follows :

1) Students value learning English as their first foreign language. Across all faculties, English is almost universally perceived as the primary foreign language. A thematic analysis of open-ended responses revealed several key motivations:
   - English as a global language. Students often emphasized the global status of English, linking it to international mobility, globalized business environments, and universal access to information.
   - Career orientation. The English language is perceived as essential for employability, international internships, and access to multinational companies.
   - Easy to learn. Students (not many, but this point of view exists) noted the easy availability of learning materials.
   - Daily use. Respondents affirm that English is an integral part of their university life; the knowledge of English allows them to understand the crux of facts and information that they have to know as that is available only in English-language sources (e.g., media, internet, study materials).
2) Students value learning of second foreign languages (German, Spanish, French, Russian, Italian). Their motivations for learning a second foreign language are united primarily around career and personal factors:
   - Career opportunities. Students prefer learning German due to the strong presence of German businesses in the Slovak region; students perceived French and Russian as valuable for careers in diplomacy.

- Personal development and cultural interest. Curiosity about other cultures and travel opportunities were strong "drivers," particularly for learning Spanish and Italian.
- Mandatory curriculum component. Respondents indicated that university requirements guided their choice.

3) Students value learning of third foreign languages for the following reasons:
   - Students chose the third language out of personal interest or love for language learning.
   - Career aspirations. A third language was seen as a competitive advantage in the job market.
   - Easy to learn. This was based on similarity to previously learned languages.

Consequently, students of economics in Slovakia hold overwhelmingly positive perceptions of learning foreign languages, with clear practical, professional, and personal motivations. While English remains dominant, growing interest in other European languages (i.e., Italian, Spanish, French, etc.) reflects students' recognition of a multilingual world. The qualitative insights gathered offer valuable guidance for curriculum development, with the potential to shape more targeted, flexible, and student-centered language education policies.

3.3.3 The pedagogical strategy for using AI-generated content in class for language education of students of economics

Given the survey results, it is clear that a higher knowledge of foreign languages gives job-seekers advantages. However, the questionnaire analysis and synthesis revealed weaknesses in the language education of students of economics. According to the analysis, insufficient attention is given to training specialized vocabulary (the confirmation data is provided in Tables 3.5–3.7). Moreover, such educational and learning dimensions as the ability to understand speech, the improvement of intercultural skills (the confirmation data is provided in Tables 3.5–3.7), the development of translation competence (the confirmation data is provided in Tables 3.5–3.7), enhancement of writing skills and speaking skills, deepening knowledge of the culture,

Table 3.5 Evaluation of how well the topics that students study and discuss in foreign language classes prepare them for their future professional careers and contribute to the development of their knowledge and skills (the assessment of the significance in the order of importance by way of an eight-point Likert scale from 1 [extremely relevant] to 8 [not at all relevant])

	1	2	3	4	5	6	7	8
General vocabulary	201	54	27	16	8	8	8	12
Specialized vocabulary	60	71	81	61	26	14	16	5
Speaking skills	149	67	44	24	20	12	7	10
Writing skills	69	79	74	56	18	22	10	6
Negotiation skills	62	61	71	56	32	22	17	11
Intercultural skills	44	49	68	62	44	37	21	8
Translation skills	41	67	67	68	35	18	25	10
Knowledge of the history, geography, the culture of the country of the studied language	23	45	42	62	50	40	45	27

Table 3.6 How difficult it is to acquire knowledge and skills to improve the level of foreign languages (the assessment of the significance in the order of importance by way of a nine-point Likert scale from 1 [extremely difficult] to 9 [not at all difficult])

	1	2	3	4	5	6	7	8	9
General vocabulary	100	58	60	31	28	21	18	11	5
Specialized vocabulary	6	31	45	62	43	48	36	34	27
Ability to understand speech	23	46	64	64	35	41	33	17	9
Ability to understand the written text	36	77	72	50	39	21	19	12	5
Intercultural skills	14	33	38	56	60	31	44	30	22
Translation competence	11	29	53	58	61	44	32	23	20
Writing skills	17	49	63	66	50	43	25	11	8
Speaking skills	30	44	60	49	46	39	29	22	12
Knowledge of the culture, history, the political situation in the country of the studied language	12	32	51	56	39	37	47	28	28

Table 3.7 What competencies and skills do you consider very well-formed? (the assessment of the significance in the order of importance by way of a nine-point Likert scale from 1 [extremely well-formed] to 9 [not well-formed])

	1	2	3	4	5	6	7	8	9
General vocabulary	127	89	46	24	13	7	11	9	9
Specialized vocabulary	10	28	82	78	52	30	28	13	12
Ability to understand speech	37	79	85	44	32	20	17	9	10
Ability to understand the written text	70	111	56	38	17	10	15	9	7
Intercultural skills	17	25	37	69	39	51	47	26	21
Translation competence	12	44	74	72	42	32	26	20	10
Writing skills	31	50	74	66	42	32	23	8	7
Speaking skills	36	64	66	56	41	27	31	6	7
Knowledge of the culture, history, the political situation in the country of the studied language	18	32	42	60	46	42	34	38	22

history, understanding of the political situation in the country of the studied language do not receive nearly enough attention in language classes (the confirmation data is provided in Table 3.6). Knowledge of the history, geography, and culture of the country of the studied language and negotiation skills are not given due attention (the confirmation data is provided in Table 3.5). Grammar, as it turned out, is a weak point in language education (the confirmation data is provided in Figure 3.27).

The results showed that career opportunities and personal improvement (development, interests)—are the key reasons why students of economics learn foreign languages (the confirmation data is provided in Figures 3.22–3.24). Increased employability is the main benefit of knowledge of foreign languages (the confirmation data is provided in Figure 3.3.5). Pursuing an international career is a primary plan for using the knowledge of foreign languages in professional activity (the confirmation data is provided in Figure 3.26).

The results obtained force us, the researchers, to think about how to handle the revealed weaknesses. The use of AI technology, in our view, can reduce weaknesses. In this regard, it is important to consider that concerns about

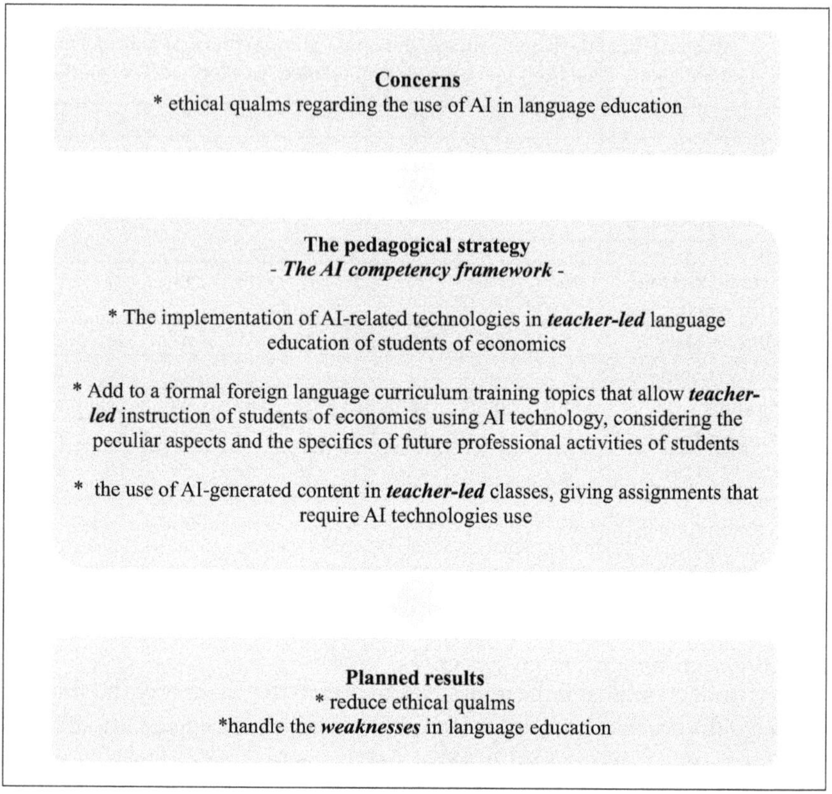

CHAPTER 3

Concerns
* ethical qualms regarding the use of AI in language education

The pedagogical strategy
- The AI competency framework -

* The implementation of AI-related technologies in *teacher-led* language education of students of economics

* Add to a formal foreign language curriculum training topics that allow *teacher-led* instruction of students of economics using AI technology, considering the peculiar aspects and the specifics of future professional activities of students

* the use of AI-generated content in *teacher-led* classes, giving assignments that require AI technologies use

Planned results
* reduce ethical qualms
*handle the *weaknesses* in language education

Figure 3.28 The pedagogical (instructional) strategy for the implementation of AI-related technologies in teacher-led language education of students of economics

using AI technology exist. And these are ethical concerns. This is according to the study performed recently (Shumeiko & Osadcha, 2024). Research findings revealed that 50% of those surveyed are concerned about ethical issues related to AI. Students expressed their opinions in the context of reflection on the process of learning EFL. Taking into consideration this important aspect, even to say, so-called ethical qualms regarding AI technology, that is the widely held view, we offer the pedagogical strategies for the implementation of AI-related technologies in teacher-led language education of students of economics (hereinafter referred to as the "strategy") (Figure 3.28).

The concept and the idea of the creation of pedagogical strategies are applied in pedagogy and allow the systematization of pedagogical ideas and propositions. The embodiment of the idea was already presented as a pedagogical strategy for IT majors in a recent co-authored publication (Osadcha & Shumeiko, 2024). The strategies (Figure 3.28) in the current research are unquestionably different. It is because those are created for teachers and students of another field of study—economics—for teacher-led language education with the implementation of AI technology. But, still, in the current research, we have not started from scratch. Essential elements of the strategies are (1) the implementation of AI-related technologies in teacher-led language education of students of economics; (2) adding to a formal foreign language curriculum training topics that allow teacher-led instruction of students of economics using AI technology, considering the peculiar aspects and the specifics of future professional activities of students; and (3) the use of AI-generated content in teacher-led classes, giving assignments that require AI technologies use.

3.3.4 Recommendations

Setting the task to *handle the weaknesses* in teacher-led language education of students of economics, considering the essence of the strategies, and having in mind the existence of the coding approach (Crompton et al., 2024), we believe that the following are needed:

1. Application of AI-powered apps for improving knowledge of specialized vocabulary

A teacher-led way of learning vocabulary with AI technologies applications is recommended to handle the weakness of insufficient attention to training specialized vocabulary (the confirmation data is provided in Tables 3.5–3.7). Vocabulary is the only subskill (Figure 3.29) in axial codes for reading (Figure 3.4).

Moreover, vocabulary learning is one of two subskills (Figure 3.30) in axial codes for reading and writing (Figure 3.3).

The knowledge of vocabulary makes it possible to read properly and write correctly. In a business context, a good knowledge of words and specialized vocabulary enables free reading and writing. The vocabulary learning process flows more easily if it is equipped with AI technologies.

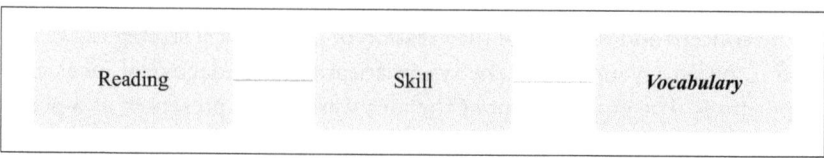

Figure 3.29 Subskill—vocabulary—in Axial codes for reading (Source: Crompton et al., 2024)

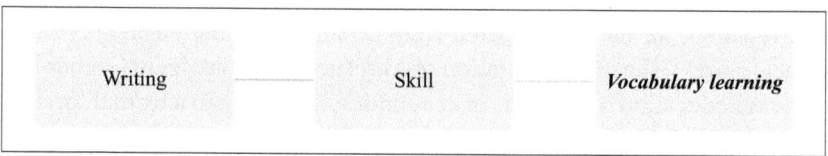

Figure 3.30 Subskill—vocabulary learning—in Axial codes for writing (Source: Crompton et al., 2024)

AI-powered apps are relevant for teacher-led classes. The following AI-powered apps and tools exist today to help you learn professional terminology. To help students acquire the terms and specialized vocabulary, "Speechify" (Figure 3.8) exploits text-to-speech technology can be used. Vocablet (<vocablet.io>) is an AI-powered vocabulary learning platform with an interactive flashcard system. It helps in mastering specialized vocabulary. Vocab-Expander (<vocab-expander.com>) enables the creation of vocabulary. As Michael Färber and Nicholas Popovic (2023) stated, the system of Vocab-Expander exploits "an ensemble of state-of-the-art word embedding techniques to suggest related terms for already given terms. In addition to word embedding models based on web text, the system also incorporates embeddings based on ConceptNet, a common-sense knowledge base." Vocab-Expander offers an easy-to-use interface that quickly confirms or rejects suggestions for terms. The created vocabulary can be sorted into a tabular form in a list view (Figure 3.31) or presented graphically in a graph form.

Table 3.8 AI tutorial systems for grammar correction

Grammarly	
Link	<www.grammarly.com>
Characteristic	It is a writing assistant that checks grammar, punctuation and spelling. It provides suggestions for improving vocabulary.
Peculiarities	Offers real-time feedback,
ELSA Speak (Figures 3.2.4–3.2.7)	
Link	<www.elsaspeak.com>
Characteristic	Pronunciation and grammar improvement for English learners
Peculiarities	Speech recognition of learners; AI technology detects grammar error in written and spoken English; provides feedback
ChatGPT (OpenAI)	
Link	<www.eopenai.com/chatgpt>
Characteristic	A conversational AI that can act as a grammar tutor by correcting grammar errors, generating exercises, and providing explanations
Peculiarities	Generating exercises by user's queries

2. Application of AI-powered tools for improving grammar proficiency

Grammar is one of two subskills (Figure 3.32) in axial codes for writing (Figure 3.3).

To handle the weakness of insufficient attention to grammar training (the confirmation data is provided in Figure 3.27), a teacher-led way of learning grammar with AI technologies applications, such as Grammarly, ELSA Speak, ChatGPT (OpenAI), is recommended (Table 3.8).

3. Application of AI-powered tools for translation purposes

Translation is one of four subskills (Figure 3.33) in axial codes for writing (Figure 3.3).

Moreover, digital translation is one of three subskills (Figure 3.34) in axial codes for reading (Figure 3.4).

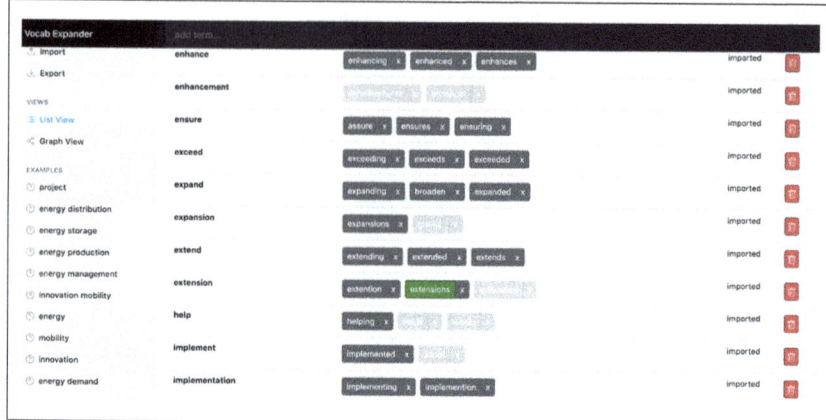

Figure 3.31 Vocab-Expander: specialized vocabulary in tabular form (Source: <vocab

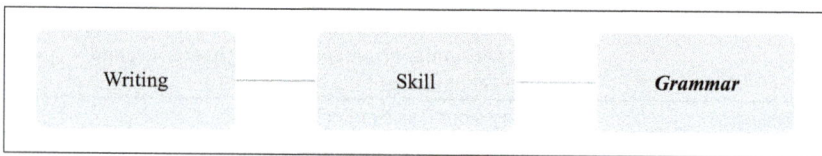

Figure 3.32 Subskill—grammar—in Axial codes for writing (Source: Crompton et al., 2024)

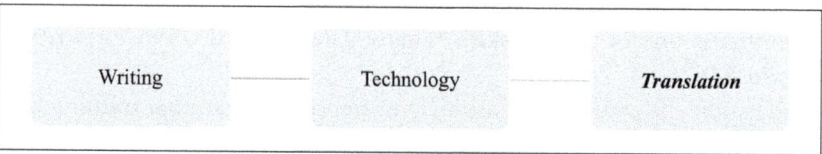

Figure 3.33 Subskill—translation—in Axial codes for writing (Source: Crompton et al., 2024)

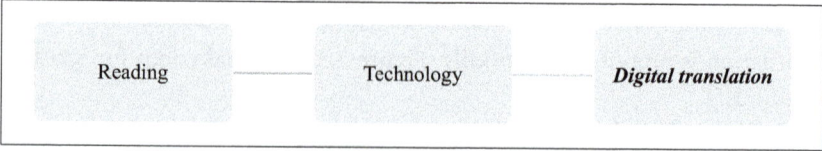

Figure 3.34 Subskill—digital translation—in Axial codes for reading (Source: Crompton et al., 2024)

-expander.com>)

To handle the weakness of insufficient attention to the development of the translation competence (the confirmation data is provided in Figures 3.21, 3.22, and 3.23), a teacher-led way of developing this competence by applying AI technologies, such as Google Translate, DeepL Translate, Microsoft Translator, Riverso Content, ChatGPT (OpenAI), is recommended.

4. Application of AI-powered tools for <u>enhancing speaking</u>

Good language proficiency opens doors to global opportunities. To handle the weakness of insufficient attention to the development of speaking skills, a teacher-led way of developing this speaking by applying AI technologies, such as ELSA Speak, Leya, and Loora, is recommended.

5. Application of AI-powered tools for <u>enhancing cultural awareness and political savvy</u>

To handle the weakness of insufficient attention to the improvement of intercultural skills (the confirmation data is provided in Tables 3.5–3.7), deepening knowledge of the culture, history, understanding of the political situation in the country of the studied language do not receive nearly enough attention in language classes (the confirmation data is provided in Table 3.6); Knowledge of the history, geography, the culture of the country of the studied language, and negotiation skills are not given due attention (the confirmation data is provided in Table 3.5), AI resources for teachers, such as CATHOVEN Language Hub (Figure 3.17–3.20), is recommended.

3.3.4 Directions for future research

The practical application of AI-powered resources in language education for students of economics is the main direction that can be contributed to future research

Bibliography

Abro, A. A., Talpur, M. S. H., & Jumani, A. K. (2023). Natural language processing challenges and issues: A literature review. Gazi Univ. J. Sci., 36(4), 1522–1536. <https://doi.org/10.35378/gujs.1032517>.

Barrot, J. S. (2021). Using automated written corrective feedback in the writing classrooms: effects on L2 writing accuracy. Computer Assisted

Language Learning, 36(4), 584–607. <https://doi.org/10.1080/09588221.2021.1936071>.

Bell, A., Davies, B., & Ammari, H. (2022). Bernhard Riemann, the ear, and an atom of consciousness. Found Sci., 27, 855–873. <https://doi.org/10.1007/s10699-021-09813-1>.

Bibauw, S., François, T., & Desmet, P. (2019). Discussing with a computer to practice a foreign language: Research synthesis and conceptual framework of dialogue-based CALL. Computer Assisted Language Learning, 32(8), 1–51.

Chen, Y., Wang, H., Yu, K., & Zhou, R. (2024). Artificial intelligence methods in natural language processing: a comprehensive review. Highlights in Science, Engineering and Technology, 85, 545–550. <https://doi.org/10.54097/vfwgas09>.

Coniam, D. (2014). The linguistic accuracy of chatbots: Usability from an ESL perspective. Text & Talk, 35(5), 545–567.

Crompton, H., Edmett, A., Ichaporia, N., & Burke, D. (2024). AI and English language teaching: Affordances and challenges. British Journal of Educational Technology, 55, 2503–2529. <https://doi.org/10.1111/bjet.13460>.

Crosthwaite, P., & Steeples, B. (2022). Data-driven learning with younger learners: Exploring corpus-assisted development of the passive voice for science writing with female secondary school students. Computer Assisted Language Learning. Advance online publication. <https://doi.org/10.1080/09588221.2022.2068615>.

Daniels, P., & Iwago, K. (2017). The suitability of cloudbased speech recognition engines for language learning. The JALT CALL Journal, 13(3), 229–239. <https://doi.org/10.29140/jaltcall.v13n3.220>.

Dizon, G., & Gayed, J. (2021). Examining the impact of Grammarly on the quality of mobile L2 writing. The JALT CALL Journal, 17(2), 74–92.

Dizon, G. & Gold, J. (2023). Exploring the effects of Grammarly on EFL students' foreign language anxiety and learner autonomy. The JALT CALL Journal, 19(3). <https://doi.org/10.29140/jaltcall.v19n3.1049>.

Ebadi, S., & Saeedian, A. (2016). The effects of computerized dynamic assessment on promoting at-risk advanced Iranian EFL students' reading skills. Issues in Language Teaching, 4(2), 1–26. <https://doi.org/10.22054/ilt.2015.7224>.

Edmett A., Ichaporia N., Cromptom H., & Crichton R. (2023). Artificial intelligence and English language teaching: preparing for the future. British Council. <http://doi.org/10.57884/78EA-3C69>.

Evers, K., & Chen, S. (2020). Effects of an automatic speech recognition system with peer feedback on pronunciation instruction for adults. Computer Assisted Language Learning, 35(7), 1–21. <https://doi.org/10.1080/09588221.2020.1839504>.

Färber & Popovic. (2023). Vocab-Expander: a system for creating domain-specific vocabularies based on word embeddings. Conference: Recent Advances in Natural Language Processing (RANLP'23). August 2023. <https://doi.org/10.26615/978-954-452-092-2_037>.

Fryer, L. K., Coniam, D., Carpenter, R., & Lăpușneanu, D. (2020). Bots for language learning now: current and future directions. Language, Leaning and Technology, 24(2), 8–22. <http://hdl.handle.net/10125/44719>.

Fu, Q. K., Zou, D., Xie, H., & Cheng, G. (2022). A review of AWE feedback: Types, learning outcomes, and implications. Computer Assisted Language Learning, Advanced online publication. <https://doi.org/10.1080/09588221.2022.2033787>.

Golonka, E. M., Bowles, A. R., Frank, V. M., Richardson, D. L., & Freynik, S. (2012). Technologies for foreign language learning: a review of technology types and their effectiveness. Computer Assisted Language Learning, 27(1), 70–105. <https://doi.org/10.1080/09588221.2012.700315>.

Godwin-Jones, R. (2021). Big data and language learning: Opportunities and challenges. Language, Learning and Technology, 25(1), 4–19.

Godwin-Jones, R. (2022). Partnering with AI: Intelligent writing assistance and instructed language learning. Language, Learning and Technology, 26(2), 5–24.

Hadley, G., & Charles, M. (2017). Enhancing extensive reading with data-driven learning. Language, Learning and Technology, 21(3), 131–152.

Hew, K. F., Huang, W., Du, J., et al. (2023). Using chatbots to support student goal setting and social presence in fully online activities: learner engagement and perceptions. J Comput High Educ., 35, 40–68. <https://doi.org/10.1007/s12528-022-09338-x>.

Hwang, G. J., & Chang, C. Y. (2021). A review of opportunities and challenges of chatbots in education. Interactive Learning Environments, 31(7), 4099–4112. <https://doi.org/10.1080/10494820.2021.1952615>.

Illingworth, S. (2023). ChatGPT: Students could use AI to cheat, but it's a chance to rethink assessment altogether. The Conversation. <https://theconversation.com/chatgpt-students-could-use-ai-to-cheatbut-its-a-chance-to-rethink-assessment-altogether-198019>.

Jeon, J. (2022). Exploring AI chatbot affordances in the EFL classroom: young learners' experiences and perspectives. Computer Assisted Language Learning, 37(1–2), 1–26. <https://doi.org/10.1080/09588221.2021.2021241>.

Jinming, D., & Ben Kei, D. (2024). Transforming language education: A systematic review of AI-powered chatbots for English as a foreign language speaking practice. Computers and Education: Artificial Intelligence, 6, 100230. <https://doi.org/10.1016/j.caeai.2024.100230>.

Kamrood, A. M., Davoudi, M., Ghaniabadi, S., & Amirian, S. M. R. (2021). Diagnosing L2 learners' development through online computerized dynamic assessment. Computer Assisted Language Learning, 34(7), 868–897. <https://doi.org/10.1080/09588221.2019.1645181>.

Kargar Behbahani, H., & Karimpour, S. (2024). The impact of computerized dynamic assessment on the explicit and implicit knowledge of grammar. Computer Assisted Language Learning, 1–22. <https://doi.org/10.1080/09588221.2024.2315504>.

Kabudi, T., Ilias Pappas, & Dag Håkon Olsen. (2021). AI-enabled adaptive learning systems: A systematic mapping of the literature. Computers and Education: Artificial Intelligence, 2, 100017. <https://doi.org/10.1016/j.caeai.2021.100017>.

Khurana, D., Koli, A., Khatter, K., et al. (2023). Natural language processing: state of the art, current trends and challenges. Multimed Tools Appl., 82, 3713–3744. <https://doi.org/10.1007/s11042-022-13428-4>.

Kazu, I. Y. & Kuvvetli, M. (2023). The influence of pronunciation education via artificial intelligence technology on vocabulary acquisition in learning English. International Journal of Psychology and Educational Studies, 10(2), 480–493. <https://dx.doi.org/10.52380/ijpes.2023.10.2.1044>.

Kim, H.-S., Kim, N. Y., & Cha, Y. (2021). Is it beneficial to use AI chatbots to improve learners' speaking performance? The Journal of Asia TEFL, 18(1), 161–178.

Koltovskaia S. (2023). Postsecondary L2 writing teachers' use and perceptions of Grammarly as a complement to their feedback. ReCALL, 35(3), 290–304. <https://doi.org/10.1017/S0958344022000179>.

Kushmar, L. V., Vornachev, A. O., Korobova, I. O., & Kaida, N. O. (2022). Artificial intelligence in language learning: What are we afraid of? Arab World English Journal (AWEJ) Special Issue on CALL, 8, 262–273. <https://dx.doi.org/10.24093/awej/call8.18>.

Lee, D., Kim, H-h., & Sung, S-H. (2023). Development research on an AI English learning support system to facilitate learner-generated-context-based learning. Education Tech Research Dev., 71, 629–666. <https://doi.org/10.1007/s11423-022-10172-2>.

Liu, S.-C. & Hung, P.-Y. (2016). Teaching pronunciation with computer assisted pronunciation instruction in a technological university. Universal Journal of Educational Research, 4(9), 1939–1943. <https://doi.org/10.13189/ujer.2016.040902>.

Long, M. H. (1996). The role of the linguistic environment in second language acquisition. In W. C. Ritchie, & T. K. Bhatia (Eds.), Handbook of second language acquisition (pp. 413–468). New York: Academic Press.

Liu, D., Bridgeman, A., & Miller, B. (2023). As uni goes back, here's how teachers and students can use ChatGPT to save time and improve learning. The Conversation. <https://theconversation.com/as-unigoes-back-heres-how-teachers-and-students-can-use-chatgpt-to-save-time-and-improve-learning199884>.

Loble, L. (2023). The rise of ChatGPT shows why we need a clearer approach to technology in schools. The Conversation. <https://theconversation.com/the-rise-of-chatgpt-shows-why-we-need-a-clearerapproach-to-technology-in-schools-199596>.

Mohamed, V., Adiel, M., Khanan, A., & Elsadig, M. A. (2024). The impact of artificial intelligence on language translation: a review. IEEE Access, 12, 25553–25579. <https://doi.org/10.1109/ACCESS.2024.3366802>.

Peng., Z., Wang, X., Han, Q., Zhu, J., Ma, X., & Qu, H. (2023). Storyfier: exploring vocabulary learning support with text generation models. Conference: UIST'23: The 36th Annual ACM Symposium on User

Interface Software and Technology. <https://doi.org/10.1145/3586183.3606786>.

Pérez-Paredes, P., Guillamón, C. O., Vyver, J. V., Meurice, A., Jiménez, P. A., Conole, G., & Hernándezd, P. S. (2019). Mobile data-driven language learning: affordances and learners' perception. System, 84, 145–159.

Pérez-Paredes, P. (2022). A systematic review of the uses and spread of corpora and data-driven learning in CALL research during 2011–2015. Computer Assisted Language Learning, 35(1–2), 36–61. <https://doi.org/10.1080/09588221.2019.1667832>.

Pishghadam, R., & Barabadi, E. (2012). Constructing and validating computerized dynamic assessment of L2 reading comprehension. IJAL, 15(1), 73–95. <http://ijal.khu.ac.ir/article-1-79-en.html>.

Osadcha, K., & Shumeiko, N. (2024). Artificial intelligence, the labor market, and education for sustainable development: the points of intersection. IOP Conf. Series: Earth and Environmental Science, 1415, 012015. IOP Publishing. <https://doi.org/10.1088/1755-1315/1415/1/012015>.

Qin, T., & Zhang, J. (2019). Computerized dynamic assessment and second language learning: Programmed mediation to promote future development. Journal of Cognitive Education and Psychology, 17(2), 198–213. <https://doi.org/10.1891/1945-8959.17.2.198>.

Randall, T. S., & Urbanski, K. (2023). Development of a computerized dynamic assessment program for second language grammar instruction and assessment. Language and Sociocultural Theory, 10(1). <https://doi.org/10.1558/lst.21006>.

Riemann, B. (1900). On psychology and metaphysics: being the philosophical fragments of Bernhard Riemann (translation by C. J. Keyser from Riemann's Gesammelte Mathematische Werke). The Monist, 10, 198–215. <https://doi.org/10.5840/monist19001029>

Riemann, B. (1984). The mechanism of the ear (translation of Mechanik des Ohres, 1866, by David Cherry, Robert Gallagher, and John Sigerson). Fusion, 6(3), 31–38.

Riemann, B. (1990). Gesammelte mathematische werke. In H. Weber (Ed.), Collected papers. Berlin: Springer.

Riemann, B. (1995). Philosophical fragments (translation by D. Cherry from Riemann's Gesammelte mathematische Werke). 21st Century Science & Technology, 8(4), 50–62.

Ritchey, T. (1991). On scientific method: based on a study by Bernhard Riemann. Systems Research, 8(4), 21–41.

Rowe, L. W. (2022). Google Translate and biliterate composing: second-graders' use of digital translation tools to support bilingual writing. TESOL Quarterly, 56(3), 883–905. <https://dx.doi.org/10.1002/tesq.3143>.

Shumeiko, N. (2024). Exploiting the potential of artificial intelligence in teaching English: Some scientific aspects. In Proceedings of XXXVII DidMatTech 2024 Conference: New Methods and Technologies in Education, Research and Practice (pp. 46–57). J. Selye University - Komárno, Trnava University in Trnava. <http://didmattech.ujs.sk/proceeding>.

Shumeiko, N., & Osadcha, K. (2024). Application of artificial intelligence in higher education institutions for developing soft skills of future specialists in the sphere of information technology. Journal of Physics: Conference Series 2871, 012027. IOP Publishing. <https://doi.org/10.1088/1742-6596/2871/1/012027>.

Shumeiko, N., & Spišiaková, M. (2025). Acquisition of English as a Foreign Language and Resources for Teaching in a Time of Generative Artificial Intelligence (AI) (pp. 1-40). Chapter 1. In: Khouya, Y. B., Alaoui, A. I. (eds), Application of AI in the Teaching and Learning of English as a Foreign Language (EFL), p. 500. IGI Global, USA.< https://www.igi-global.com/book/application-teaching-learning-english-foreign/361715>

Son, J.-B., Ružić, N. K., & Philpott, A. (2023). Artificial intelligence technologies and applications for language learning and teaching. J. China Comput. Assist. Lang. Learn. <https://doi.org/10.1515/jccall-2023-0015>.

Smutny, P., & Schreiberova, P. (2020). Chatbots for learning: a review of educational chatbots for the Facebook Messenger. Computers & Education, 151, 103862.

Tajeddin, Z., & Griffiths, C. (Eds.). (2023). Language education programs: Perspectives on policies and practices. Springer Cham. <https://doi.org/10.1007/978-3-031-38754-8>

Tono, Y., Satake, Y., & Miura, A. (2014). The effects of using corpora on revision tasks in L2 writing with coded error feedback. ReCALL, 26(2), 147–162.

Thompson, A., Gallacher, A. & Howarth, M. (2018). Stimulating task interest: Human partners or chatbots? In P. Taalas, J. Jalkanen, L. Bradley & S. Thouësny (Eds.), Future-proof CALL: Language learning as exploration and encounters—short papers from EUROCALL 2018 (pp. 302–306). <https://doi.org/10.14705/rpnet.2018.26.854>.

Vincent, J. (2022). AI-generated answers temporarily banned on coding Q&A site stack overflow. Verge. <https://www.theverge.com/2022/12/5/23493932/chatgpt-ai-generated-answers-temporarilybanned-stack-overflow-llms-dangers>.

Wang, J., Hwang, G.-W., & Chang, C.-Y. (2021). Directions of the 100 most cited chatbot-related human behavior research: A review of academic publications. Computers and Education: Artificial Intelligence, 2, 100023.

Wang, X., Pang, H., Wallace, M. P., Wang, Q., & Chen, W. (2022). Learners' perceived AI presences in AI-supported language learning: A study of AI as a humanized agent from community of inquiry. Computer Assisted Language Learning. Advance publication. <https://doi.org/10.1080/09588221.2022.2056203>.

Wu, Y.-j. (2021). Discovering collocations via data-driven learning in L2 writing. Language, Learning and Technology, 25(2), 192–214.

Zhai, X. (2022). ChatGPT user experience: implications for education. Social Science Research Network (SSRN). <https://doi.org/10.2139/ssrn.4312418>.

Zhai, N., & Ma, X. (2022). Automated writing evaluation (AWE) feedback: a systematic investigation of college students' acceptance. Computer Assisted Language Learning, 35(9), 2817–2842.

Conclusion

This monograph, entitled "Why Artificial Intelligence? AI Tools in Information Technology and Foreign Language Education at the Tertiary Level," has sought to provide a comprehensive analysis of the role of AI in tertiary education. The research work focused particularly on AI tools and their use in information technology and foreign language education. The monograph findings confirm that the use of AI in tertiary education has the potential to enhance students' competencies, improve knowledge and skills, and enrich the teaching–learning process. The findings of the monograph highlight the possibilities for using AI not as a replacement for traditional pedagogy, but as a supportive and innovative tool that fosters learning engagement and efficiency.

Taken together, the thoughts, views, and considerations gathered in this study address two fundamental research questions. The first addressed the topic of how tertiary education can leverage the potential of AI to enhance the knowledge of IT students and economics students. And, correspondingly, the second concentrated on whether AI can serve as a reasonable means of education for IT students and for economics students learning foreign languages.

To answer these questions, considering the objectives of the research - to conduct an analysis of existing scientific knowledge about AI through bibliometric analysis, to suggest pedagogically checked approaches for integrating AI in the tertiary level education, and to provide examples of applying AI in

IT and language classrooms. These objectives were developed across three chapters, each offering theoretical perspectives as well as practical suggestions, illustrations, and learning and teaching materials.

Chapter 1 presented a bibliometric analysis, highlighting the rapid growth of research publications on AI in education. This finding reflects a global trend toward recognizing AI as a transformative potential for education. Chapter 2 explored the application of Microsoft Copilot. In this chapter, it is illustrated how the AI-powered resource, Microsoft Copilot, can support the training of IT majors, particularly in solving complex management and technical tasks. Chapter 3 examined the integration of AI tools into foreign language teaching at the tertiary level. The research focused on tools such as Speechify (a text-to-speech technology), ELSA, Leya, Loora, and the CATHOVEN Language Hub, all of which aid university lecturers and students in improving language acquisition through accessible and engaging modes. The results presented here support the view that AI tools hold potential for both IT education and foreign language learning.

In conclusion, the research has made contributions to both theoretical understandings and practical implications in the field of AI in tertiary education. By combining bibliometric evidence, pedagogical considerations, thoughts, and practical examples, this monograph offers a balanced perspective on the promises and challenges of AI integration. The outcomes of this study provide an exploration of innovative AI opportunities in teaching and learning. The target audience for this monograph is scholars and university lecturers who seek to responsibly integrate AI into their university classes, aiming to foster a future-oriented academic environment that incorporates AI.

Appendix A

Questions/answers for the multiple-choice task (Figure 3.2.10) generated by CATHOVEN Language Hub (Analysis by <nexthub.cathoven.com>)]

Untitled Text

Read the passage and choose the correct answer.

Passage

To Sherlock Holmes she is always the woman. I have seldom heard him mention her under any other name. In his eyes she eclipses and predominates the whole of her sex. It was not that he felt any emotion akin to love for Irene Adler. All emotions, and that one particularly, were abhorrent to his cold, precise but admirably balanced mind. He was, I take it, the most perfect reasoning and observing machine that the world has seen, but as a lover he would have placed himself in a false position. He never spoke of the softer passions, save with a gibe and a sneer. They were admirable things for the observer—excellent for drawing the veil from men's motives and actions. But for the trained reasoner to admit such intrusions into his own delicate and finely adjusted temperament was to introduce a distracting factor which might throw a doubt upon all his mental results. Grit in a sensitive instrument, or a crack in one of his own high-power lenses, would not be more disturbing than a strong emotion in a nature such as his. And yet there was but one woman to him, and that woman was the late Irene Adler, of dubious and questionable memory.

Questions

1. What is the primary reason Sherlock Holmes admires Irene Adler?

A. She is unique and dominates over others in his perception.
B. He feels a romantic affection for her.
C. Her memory is always a source of emotional comfort for him.
D. She is a representation of all women in his life.

Answer: A

Explanation: Sherlock Holmes admires Irene Adler because he perceives her as singular and superior compared to other women.

Answer location: "In his eyes she eclipses and predominates the whole of her sex."

2. How does Sherlock Holmes view emotions in relation to his work as a detective?

A. They are irrelevant and do not impact his work.
B. They should be embraced to improve his analyses.
C. They are essential for understanding human motives.
D. They are a threat to his reasoning and observational skills.

Answer: D

Explanation: Holmes sees emotions as a potentially destabilizing influence that could impair his analytical capabilities.

Answer location: "But for the trained reasoner to admit such intrusions into his own delicate and finely adjusted temperament was to introduce a distracting factor which might throw a doubt upon all his mental results."

3. Why does Sherlock Holmes seldom express emotions of love or affection?

A. He is not capable of understanding such emotions.
B. He believes love is the most vital emotion.
C. He considers them abhorrent and as potential distractions.
D. He finds such emotions to be admirable and beneficial.

Answer: C

Explanation: Holmes regards emotions, especially love, as undesirable and a threat to his mental equilibrium.

Answer location: "All emotions, and that one particularly, were abhorrent to his cold, precise but admirably balanced mind."

4. In what way does Sherlock Holmes describe his reasoning capabilities?

A. As the most perfect reasoning and observing machine.
B. As something inferior that needs improvement.
C. As unique but unreliable.
D. As an emotionally driven process.

Answer: A

Explanation: Holmes views his reasoning capabilities as unparalleled, akin to a perfect machine.

Answer location: "He was, I take it, the most perfect reasoning and observing machine that the world has seen."

Appendix B

Questions for a True/False/Not given task (Figure 3.2.12) generated by CATHOVEN Language Hub (Analysis by <nexthub.cathoven.com>)

Untitled Text

Read the passage and mark the statements True, False or Not Given.
True = the statement matches the information in the passage.
False = the statement contradicts the information in the passage.
Not Given = the information is not found in the passage.

Passage

To Sherlock Holmes she is always the woman. I have seldom heard him mention her under any other name. In his eyes she eclipses and predominates the whole of her sex. It was not that he felt any emotion akin to love for Irene Adler. All emotions, and that one particularly, were abhorrent to his cold, precise but admirably balanced mind. He was, I take it, the most perfect reasoning and observing machine that the world has seen, but as a lover he would have placed himself in a false position. He never spoke of the softer passions, save with a gibe and a sneer. They were admirable things for the observer—excellent for drawing the veil from men's motives and actions. But for the trained reasoner to admit such intrusions into his own delicate and finely adjusted temperament was to introduce a distracting factor which might throw a doubt upon all his mental results. Grit in a sensitive instrument, or a crack in one of his own high-power lenses, would not be more disturbing than a strong emotion in a nature such as his. And yet there was but one woman to him, and that woman was the late Irene Adler, of dubious and questionable memory.

Questions

1. Sherlock Holmes was deeply in love with Irene Adler.

Answer: False

Explanation: El texto señala claramente que Holmes no sentía ninguna emoción semejante al amor por Irene Adler.

Answer location: "It was not that he felt any emotion akin to love for Irene Adler. All emotions, and that one particularly, were abhorrent to his cold, precise but admirably balanced mind."

2. Irene Adler is remembered fondly by Sherlock Holmes.

Answer: Not Given
Explanation: El texto no proporciona información sobre si Holmes recuerda a Adler con cariño o no.
Answer location: "N/A"

3. Strong emotions could interfere with Sherlock Holmes's reasoning abilities.

Answer: True
Explanation: Se indica que para Holmes, las emociones fuertes podrían ser un factor perturbador que afectaría sus resultados mentales.
Answer location: "But for the trained reasoner to admit such intrusions into his own delicate and finely adjusted temperament was to introduce a distracting factor which might throw a doubt upon all his mental results."

4. Irene Adler was the only woman of significance to Sherlock Holmes.

Answer: True
Explanation: El texto menciona que para Holmes, solo había una mujer que le importaba, y esa era Irene Adler.
Answer location: "And yet there was but one woman to him, and that woman was the late Irene Adler, of dubious and questionable memory."

Appendix C

Feedback for students' writing (Figure 3.2.13) generated by CATHOVEN Language Hub (Analysis by <nexthub.cathoven.com>)

APPENDICES

Revision of *Untitled Text*

revised

Revised writing

It is clear that ~~the people who~~ play games the most ~~is people~~ are those under the age of eighteen. ~~The~~ And the percentage of male players is larger than that of female players. We can also ~~can~~ see ~~the~~ that action games take up a majority portion of the total ~~of game:~~ games.

The first chart shows the ages of players. There are 39% ~~of~~ players ~~are~~ under 18, and the percentage of players ~~that~~ between 18 ~~to~~ and 36 ~~year~~ years old is slightly smaller than ~~that~~ those under 18. The number of players who ~~are~~ over 36 years old ~~are~~ is the ~~least~~ smallest of all ~~payers~~ players in Korea. ~~And the~~
~~The~~ second chart compares the gender of players. The number of male players is almost three times ~~as much as~~ that of female players.

Revisions

Revised
It is clear that ~~the people who~~ play games the most ~~is people~~ are those under the age of eighteen.

Revised
The ~~And the~~ percentage of male players is larger than that of female players.

Revised
We can also ~~can~~ see ~~the~~ that action games take up a majority portion of the total ~~of game:~~ games.

204

APPENDICES

Revised

There are 39% ~~of~~ players ~~are~~ under 18, and the percentage of players ~~that~~ between 18 ~~to~~ and 36 ~~year~~ years old is slightly smaller than ~~that~~ those under 18.

Revised

The number of players who ~~are~~ over 36 years old ~~are~~ is the ~~least~~ smallest of all ~~payers~~ players in Korea.

Revised

The ~~And the~~ second chart compares the gender of players.

Revised

The number of male players is almost three times ~~as much as~~ that of female players.

205

Appendix D

Questionnaire (selected questions)

No	Question	The type of question	The analysis results
1	Outline the reasons for learning English as a first foreign language	Open question	Figure 3.3.1
2	Outline the reasons for learning English as a second foreign language	Open question	Figure 3.3.3
3	Outline the reasons for learning English as a third foreign language	Open question	Figure 3.3.4
4	Prioritize importance of learning a foreign language within the context of general education	A five-point Likert scale	Chapter 3.3.1
5	Prioritize importance of learning a foreign language for professional use after graduation	A five-point Likert scale	Chapter 3.3.1

APPENDICES

No	Question	The type of question	The analysis results
6	Evaluate how the foreign language class topics are in line with your future profession and contribute to developing your knowledge and skills · General vocabulary · Specialized vocabulary · Speaking skills · Writing skills · Negotiation skills · Intercultural skills · Translation skills · Knowledge of the history, geography, the culture of the country of the studied language	An eight-point Likert scale	Table 3.5
7	Evaluate how difficult it is to acquire knowledge and skills to improve the level of foreign languages · General vocabulary · Specialized vocabulary · Ability to understand speech · Ability to understand the written text · Intercultural skills · Translation skills · Speaking skills · Writing skills · Speaking skills · Knowledge of the culture, history, the political situation in the country of the studied language	A nine-point Likert scale	Table 3.6

APPENDICES

No	Question	The type of question	The analysis results
8	What competences and skills do you consider very well-formed? · General vocabulary · Specialized vocabulary · Ability to understand speech · Ability to understand the written text · Intercultural skills · Translation skills · Speaking skills · Writing skills · Speaking skills · Knowledge of the culture, history, the political situation in the country of the studied language	A nine-point Likert scale	Table 3.7
9	Opt for a self-assessment of your knowledge of the first foreign language	A five-point Likert scale	Chapter 3.3.1
10	Opt for a self-assessment of your knowledge of the second foreign language	A five-point Likert scale	Chapter 3.3.1
11	Opt for a self-assessment of your knowledge of the third foreign language	A five-point Likert scale	Chapter 3.3.1
12	Determine the benefits from knowledge a foreign language and plans for using the knowledge in future professional activity	Open question	Figure 3.3.5, Figure 3.3.6

No	Question	The type of question	The analysis results
13	In foreign language classes teachers focus on the cultivation of linguistic skills (speaking, writing), language competence (grammar), and the ability to understand the different elements of spoken and written language (language comprehension). Which of them are developed in a greater extent and which of them are developed in a lesser extent thanks to the foreign language classes?	Open question with proposed options	Figure 3.3.7

www.ingramcontent.com/pod-product-compliance
Ingram Content Group UK Ltd.
Pitfield, Milton Keynes, MK11 3LW, UK
UKHW021834210426
5322IPUK00018B/262